Science and the Media

Science and the Media: Delgado's Brave Bulls and the Ethics of Scientific Disclosure

Peter J. Snyder
Linda C. Mayes
Dennis D. Spencer

With Invited Essays

AMSTERDAM • BOSTON • HEIDELBERG • LONDON • NEW YORK • OXFORD
PARIS • SAN DIEGO • SAN FRANCISCO • SINGAPORE • SYDNEY • TOKYO
Academic Press is an imprint of Elsevier

Academic Press is an imprint of Elsevier
32 Jamestown Road, London NW1 7BY, UK
Radarweg 29, PO Box 211, 1000 AE Amsterdam, The Netherlands
30 Corporate Drive, Suite 400, Burlington, MA 01803, USA
525 B Street, Suite 1900, San Diego, California 92101-4495, USA

First edition 2009

British Library Cataloguing in Publication Data
A catalogue record for this book is available from the British Library

Library of Congress Cataloging-in-Publication Data
A catalog record for this book is available from the Library of Congress

ISBN: 978-0-12-373679-6

For information on all Academic Press publications
visit our website at books.elsevier.com

Typeset by Charon Tec Ltd., A Macmillan Company. (www.macmillansolutions.com)

Printed and bound in the United States of America

09 10 9 8 7 6 5 4 3 2 1

For Molly Rose and Jacob Brendan Eli
my two wonderful children, and my sources of inspiration and love.
Peter J. Snyder

For my mother, Marion, in gratitude and admiration for her courage,
strength and grace in the face of all of life's surprises.
Linda C. Mayes

To mother and her common sense.
Dennis Spencer

Contents

Contributors

The Very Reverend Dr. Wesley Carr, M.A., Ph.D., K.C.V.O.
Dean Emeritus, Westminster Abbey, Church of England, Hants, UK

Declan P. Doogan, M.B., Ch.B., FRCP(Glas.), FFPM, DSc
President, Research & Development,
Amarin Corporation, London, UK;
Former Senior Vice President and Head, Worldwide Development,
Pfizer Inc.

Peter Fonagy, Ph.D., D.Psych., F.B.A.
Freud Memorial Professor of Psychoanalysis,
Director, Sub-department of Clinical Health Psychology,
University College London, London, UK;
Chief Executive,
Anna Freud Centre, London, UK;
Adjunct Professor of Psychiatry,
Baylor College of Medicine, Houston, TX, USA;
Clinical Professor,
Child Study Center, Yale University, School of Medicine,
New Haven, CT, USA

Jerome P. Kassirer, M.D.
Distinguished Professor,
Tufts University School of Medicine,
Editor-in-Chief Emeritus, *The New England Journal of Medicine*,
Boston, MA, USA

Ruth J. Katz, J.D., M.P.H.
Dean, School of Public Health and Health Services,
Walter G. Ross Professor of Health Policy,
The George Washington University, Washington, DC, USA

Aaron Klink, M.Div.
Westbrook Scholar of Theology and Medicine,
Duke University School of Divinity, Durham, NC, USA

Julius Landwirth, M.D., J.D.
Associate Director, Interdisciplinary Center for Bioethics,
Yale University, New Haven, CT, USA

Robert J. Levine, M.D.
Professor of Medicine and Lecturer in Pharmacology;
Director, Donaghue Initiative in Biomedical and
Behavioral Research Ethics;
Director, Law, Policy and Ethics Core,
Center for Interdisciplinary Research on AIDS,
Yale University, New Haven, CT, USA

Linda C. Mayes, M.D.
Arnold Gesell Professor of Child Psychiatry, Pediatrics, Psychology,
and Epidemiology and Public Health, Child Study Center,
Yale University School of Medicine,
New Haven, CT, USA

Harry W. McConnell, M.D., FRCP(C)
Director, Institute for Sustainable Health, Education and Development;
Professor of Neurology and Psychiatry,
Griffith University School of Medicine, Queensland, Australia

Ashley Pardy, M.A.
Griffith University School of Medicine,
Gold Coast Campus, Queensland, Australia

Pat Shipman, Ph.D.
Adjunct Professor,
Department of Anthropology,
The Pennsylvania State University, University Park, PA, USA

Peter J. Snyder, Ph.D.
Vice President for Clinical Research, Rhode Island Hospital,
The Miriam Hospital and Bradley Hospital,
Providence, RI, USA;
Professor, Department of Clinical Neurosciences (Neurology),
The Warren Alpert Medical School of Brown University,
Providence, RI, USA;
Adjunct Professor, Child Study Center, Yale University School of Medicine,
New Haven, CT, USA

David H. Smith, Ph.D.
Director, Interdisciplinary Center for Bioethics,
Yale University, New Haven, CT, USA

Dennis D. Spencer, M.D.
Harvey and Kate Cushing Professor of Neurosurgery,
Chair, Department of Neurosurgery, Yale University School of Medicine,
New Haven, CT, USA

Laura Spinney, B.Sc.
Freelance Science Journalist,
London, UK

John H. Warner, Ph.D.
Avalon Professor and Chair of the Section of the History of Medicine,
Yale University School of Medicine, New Haven, CT, USA

Acknowledgments

This volume is the result of the combined efforts of several colleagues and institutions whom we would like to thank profusely for their support, encouragement and assistance. Each of the essayists who contributed to this book took time away from enormously busy schedules, as well as their important responsibilities, to author chapters for this book. The very high caliber of our essayists, all of whom are senior leaders in their respective fields, has been especially gratifying as, over the course of one's career, it becomes harder to find the time and extra energy required to author book chapters. We truly appreciate the generosity, interest and good will of our essayists for this volume.

Many of these authors, from across the United States, Great Britain and Australia, came together in July of 2007, when we hosted a small conference held at the Child Study Center, Yale University School of Medicine, New Haven, Connecticut, USA. That nearly 100 individuals attended this conference, on a Friday in the summertime and with only 4 weeks of limited advance advertising, serves as evidence of the high level of interest that faculty and students alike share for the topics covered in this book. We want to thank The Frank Robinson Lectureship (to the Yale Department of Neurosurgery), and especially Mrs. Gloria Robinson, for their generous support of this conference. We also thank Ms. Carol Pollard, and the Yale Interdisciplinary Center for Bioethics, for an educational grant that provided partial funding for the meeting. Finally, we thank Dr. Charles C. Lowe, and the Department of Psychology at the University of Connecticut (Storrs, Connecticut), for their generous financial support for both the conference as well as help to defray additional costs incurred in the production of this volume. The conference, titled *"The Publicizing of Biomedical Research: The Ethical Dance Between Scientists and the Media"* (July 13, 2007; Cohen Auditorium, Yale Child Study Center) was jointly sponsored by the Department of Psychology, University of Connecticut, and the Yale University School of Medicine.

We also wish to thank Drs. Edward L. Etkind and Robert A. Feldman for their advice and support throughout this project. Several of Prof. Snyder's graduate students, most notably Colleen E. Jackson, Robert H. Pietrzak and Kate Papp, went out of their way to assist in the editing and critiquing of many portions of this book, and their advice and effort is very much appreciated. Finally, we wish to thank Dr. Johannes Menzel, our Publisher, and Ms. Kristi A.S. Gomez, our Developmental Editor – both of the Life Sciences Division of Elsevier B.V. – for their enormous amount of support throughout this project. Johannes was an early ardent supporter of the concept for this volume, and he was instrumental in the initial meetings that led to all of the decisions regarding the scope, breadth and individual essay topics/contributors. Both Johannes and Kristi are more than representatives of a large and respected scientific publishing corporation; they are our trusted colleagues.

Peter J. Snyder
Linda C. Mayes
Dennis D. Spencer
New Haven, CT, USA

CHAPTER

1

Introduction: The Ethics of Scientific Disclosure

Peter J. Snyder and Linda C. Mayes

In its day-to-day execution, science is a social enterprise, and like all enterprises involving people, science is driven as much by the hopes, dreams, and vested interests of its practitioners as it is to their rational capacities to question and understand. As humans, we are also compelled to make meaning out of our experiences, and scientists are no different. Science is about discovery, about seeing a new story amidst a series of events, experiments, or observations. Sometimes the stories scientists tell capture the imagination of the public and of other story tellers, namely journalists, because these stories are clearly so unexpected, so unusual, and so potentially world-view changing. Sometimes they capture the public and journalistic imagination because at the moment of the story's creation, it touches upon a need in society that is far greater than a rational emphasis on critical inquiry. Further, especially in our western society, we have placed considerable value and primacy on science as both the road to truth as well as better and more productive lives. In this way, too, the stories scientists tell in their books and journal reports are given considerable weight, and often accompanied by suspended critical judgment. There are moments when the interests of scientists in having their discoveries gain recognition converge with the needs and interests of the public and the journalists who bring the scientists' discoveries to the public in not altogether mutually beneficial ways.

Consider the following statement: *Long-distance prayer for infertile women leads to a dramatic increase in successful pregnancies following in vitro fertilization.* This astounding claim was published in 2001 as a peer-reviewed scientific report in the *Journal of Reproductive Medicine*, a leading, top-tier medical journal. The paper, co-authored by Prof. Rogerio A. Lobo, who was then Chairman of the Department of Obstetrics and Gynecology at Columbia University School of Medicine (New York), examined the pregnancy rates for 199 women in Seoul, South Korea, who were all undergoing *in vitro* fertilization.[1] Half of these women were, without their knowledge or consent, the objects of anonymous, remote intercessory prayer; and the authors of this study reported that whereas 50 percent of infertile women in the "experimental group" who received prayer enjoyed successful pregnancies following *in vitro* fertilization, only 24 percent of the control group – those who were not prayed for – became pregnant.

Amazingly, this publication, printed in such a highly respected medical journal, was actually not some sort of elaborate practical joke. Dr. Lobo, who held the "senior author" position on the paper, told the Reuters news service that, "Essentially, there was a doubling of the pregnancy rate in the group that was prayed for."[2] Moreover, Columbia University issued a news release to state that this remarkable study had several safeguards in place to eliminate bias, and that the study itself was designed carefully in order to eliminate a variety of sources of bias.[3] With the apparent support from a prestigious university to back the veracity of this report, it comes as no surprise that on October 02, 2001, *The New York Times* reported this extraordinary discovery to a public shaken by the terrible tragedy in New York City on September 11th of that year, less than 1 month earlier, many of whom were eager to receive such news as a sign that their faith in God was not misplaced.[4] From this point the news spread to the morning television talk shows, such as the widely viewed ABC show, *"Good Morning America,"*[5] and this story found immediate support from celebrities and supposed authorities, many of whom hailed this finding as an exciting, historic scientific development. Commenting on this entire line of investigation, the popular spirituality author and radio personality, Dr. Deepak Chopra, noted that prayer experiments are "encouraging enough that we should pursue them, because if we don't, we may have missed one of the most amazing phenomena in nature."[6]

This study, and other poorly controlled studies like it, mostly authored by individuals with personal, vested interests in the direction of the study outcomes, have been the subject of reports in such mainstream publications as *Time* magazine, *Wired* magazine, *Good Housekeeping* magazine, and countless radio and television talk shows. In the case of the 2001 study by Cha *et al.*, news agencies around the world reported on this miracle, thereby raising the hopes of unknown numbers of vulnerable, anxious, and worried infertile couples.

There have been several detailed reviews of the ethical and methodological flaws in this study.[7,8] In fact, the list of odd events associated with this study is both long and bizarre. The first author, Dr. Kwang Cha, left Columbia University soon after the publication of this report, and he has repeatedly refused to respond to any questions about the study from journalists.[9,10] The second author, Daniel Wirth, has a dubious educational background (a Master's degree in "parapsychology"), a publication track record of strange reports in low-tier journals touting the clinical benefits of faith healing, non-contact therapeutic touch (NCTT), and "psychic surgery," and a criminal record that is notable for its broad range of scams, cons, and offenses. Wirth's record of offenses, leading to both criminal charges and convictions, range from defrauding the US Social Security Administration, using the stolen identity of a deceased child to obtain a false US passport, conspiracy to commit mail fraud, conspiracy to commit bank fraud, and much, much more.[9,10] Wirth is presently serving a 5-year sentence in Federal prison, and his longtime accomplice who also received the same recent felony convictions, was found dead in his prison cell in July of 2004, from an apparent suicide.[9]

Finally, the person initially identified as the "lead author" for the paper by Columbia University, and then by ABC News and *The New York Times*, Dr. Lobo, was on the Editorial Review Board of the journal in which the paper was published. Lobo initially consented to several interviews in this identified role as the lead author, but then recanted his supposed involvement in this study, claiming later that he only provided "editorial review and assistance" with the publication. Dr. Lobo allowed himself, and his office as the Chair of the Department of Obstetrics and Gynecology at one of this Nation's most prestigious of medical schools, to lend credibility to this report – which was later discovered to have never been submitted to any Institutional Review Board for ethical review or approval, prior to the identities and photographs of 199 infertile women being sent to several faith healing groups in various countries for anonymous prayer.[9,11] Like his co-authors, Dr. Lobo has also refused to comment further on his involvement in this study, and he later retracted his name as an author when the article was re-posted to the journal's web site.

In his reviews of this entire sordid affair, Dr. Bruce Flamm, a clinical professor of Obstetrics and Gynecology at the University of California-Irvine, notes how the publication of this report had emboldened the strength and conviction of claims made by several vocal proponents of faith healing and "complimentary medicine" for the treatment of infertility.[8,9] Despite the repeated detailed reviews that list the flaws of this fraudulent report in great detail, its publication nonetheless continues to be cited by those who wish the results to be true. A recent *Google* search using the search terms "Wirth, Columbia, Prayer" resulted in over 22 400 hits. Although a majority of these web sites provide skeptical reports of the study, a substantial portion appear to accept the results of the study at face value, citing that report as clear scientific proof that intercessory prayer will lead to increased success in the treatment of infertility. Perhaps more troubling, many of these web sites seem to be designed to appear to the naïve reader as sources of credible medical advice, and any details that show who the sponsors are and whether they have any particular religious or political biases, are not overtly disclosed.

As Flamm points out in his 2004 review, "the fact that a 'miracle cure' study was deemed to be suitable for publication in a scientific journal automatically enhanced the study's credibility. Not surprisingly, the news media quickly disseminated the 'miraculous' results … serious damage has been done. The [journal] editors were informed of several of the study's flaws within weeks of its publication and yet allowed the entire study to remain on their web site for 2 years. During that time the public was never given any reason to doubt the study's validity or its miraculous claims."[9]

1 Public trust in science

What can we learn from this recent vignette? Is this an isolated, rare example of scientific reporting gone wrong? This affair has raised many questions, and many of them remain unanswered. Why would two respected and well-trained infertility specialists choose to collaborate on any level with someone with a 20+ year criminal record and a highly suspect research history? Why would Dr. Lobo allow his name to be listed as the lead author on a paper that he later claimed that he had no involvement with, other than providing editorial assistance? Why would he then accept interviews as the "lead author" soon after the publication of the study, only then to deny his involvement once investigations of study conduct, including one by the Department of Health and Human Services for ethical misconduct, were undertaken? Why would such a highly respected journal choose to publish such a bizarre and confusing study without a much more rigorous review process, and why would they refuse to retract or issue any comment regarding this study for so long?

Perhaps the most important question we could ask about this study, and others like it, is what is the cost to society when scientific results are misreported to the public? This particular case is an admittedly extreme example, as the underlying science was fraudulent and the authors were either untrustworthy or uninvolved in the research that was reported under their name. In this instance, those harmed were perhaps thousands of infertile couples who are desperate to conceive and start families, and who are searching for hope, new treatments, and who all rely on the advice of experts such as Drs. Cha and Lobo. But what about lesser examples in which the science may not be frankly false, but perhaps results are slightly embellished or over-simplified in their retelling? What are the motivating forces that might lead a scientist to take risks, or perhaps a bit of "artistic license" in conveying research results in either peer-reviewed journals or to the lay presses? Conversely, what are the potential sources of secondary gain that might lead a science writer, editor, or publisher to over-simplify a complex story, to publish unqualified conclusions too soon, or to accept a report that has not been fully checked for its veracity? And, how can ethical, responsible scientists, writers, and editors protect against the often inadvertently inaccurate description of a scientific finding, despite the best of intentions?

Over the past few centuries, our culture has held a strong faith in the positive, progressive nature of biomedical science. The informed layman has turned to the media for the most up to date information on new cures and new understanding about health and illness. Scientists, in turn, use the media to make their discoveries widely known, to stake a claim to a new discovery, and to promulgate the road to health that their discoveries may offer. There is a necessary synergy between the public's wish and need to know, the media's need for stories, and scientists need for a broad audience and stage on which to display their work and make it available for public debate. In the best of worlds, this synergy is healthy and leads to greater awareness among the public, invigorating intellectual discourse, and adaptive competition among scientists. On the other hand, sometimes the relationship between science and the media leads to misleading interpretations that may be damaging to individuals seeking help based on the news, to scientists working in similar areas, and to a field in general. When things go awry, the general trust in science is rapidly eroded, and there seems to be an emerging recent trend in many countries around the globe to "police" the accurate reporting of science in a very public manner. As one example, in recent years The *Guardian* newspaper, in London, has taken to publishing a regular feature on the reporting of "bad science" (http://www.guardian.co.uk/life/badscience/).

2 Lessons from recent history

Although there are innumerable examples in which remarkable and groundbreaking scientific discoveries are accurately conveyed in the professional and lay presses each day, there are sadly many recent and historical examples in which this proverbial train was clearly derailed off the tracks. It is our goal, in this volume, to choose several contemporary and historical vignettes to perform a postmortem examination of what went wrong, and what we can learn from past mistakes. The objectives of this book, then, are to address the following topics:

- What are the causal factors that lead to inaccuracies and misleading interpretations, in the reporting of biomedical discoveries in both the professional and lay presses?
- What are the roles and responsibilities of both scientists and the journalists that scientists interact with?
- What ethical obligations do scientists have when we convey our research findings to either the peer-review professional or lay presses?
- What are the societal and economic pressures that lead to intentionally inexact reporting of important research? Are the recent concerns regarding the integrity of biomedical research, funded by pharmaceutical or biotechnology companies and then published in peer-reviewed journals, justified?
- How do political and religious movements affect the public dissemination of scientific knowledge, and the training of new scientists?
- What are the real costs to society as a result of ethical errors in judgment when research results are withheld, conveyed inaccurately, or provided without appropriate cautionary statements for public disclosure?

- How effective is the biomedical research community at policing itself in this regard? Are the protective measures that we have in place, such as the blind peer-review system for journal manuscript submissions, as effective as we desire?

In this book, we ask the question, "What are the ethical obligations of the scientist/scholar when his or her work is conveyed to other scientists or to the larger general public?" Whose responsibility is it to make sure that new knowledge is portrayed accurately? In their reporting to the general public, how can scientists balance the optimism that naturally accompanies creativity and discovery with appropriate caution and skepticism while at the same time stimulating public interest and understanding? What social forces are in play, ranging from political and religious pressures, to individual career and economic pressures that often dissuade scientists from appropriately reporting data or providing qualifying statements made to the media?

We intend to answer these pivotal questions in three ways. First, we will review several contemporary case studies in some detail, as they show these issues to be central to the conduct of, and social acceptance of, science in the 21st century. Our case studies will be limited to ones chosen from recent advances in the clinical neurosciences, but it should be immediately apparent to the reader that the choice of topic area for research inquiry is really a matter of convenience from our perspective as authors, given our own professional training and experiences. The questions and issues that we seek to explore cut across all domains in the biomedical sciences, and perhaps across all of science – as the pursuit of science is a human activity that is subject to all the social influences and pressures that any other complex social endeavor faces. By using both historical and contemporary examples, we will illustrate the continuing need to scrutinize carefully the relationship between science and the media and make the case that especially as biomedical research becomes increasingly sophisticated technologically, the opportunities for inaccuracies and misinterpretation are ever greater.

One such contemporary example, described in detail in the next chapter, highlights the strong pressures faced by journal editors to seek, attract, and publish the most exciting of "breakthrough" discoveries in the pages of their journals. In 1987, the work of a Mexican research team made international headline news when a paper reporting two surgical case studies was published in the prestigious *The New England Journal of Medicine*. In this paper, and in a second paper published 1 year later (totaling only four patients reported), the authors described a breakthrough surgical procedure for severely disabled Parkinson's disease (PD) patients, in which adrenal medullary tissue is removed from the patient's own adrenal gland (in the abdomen) and then autotransplanted into the middle of the caudate nucleus, a structure that is buried deep within the brain. For these four individual patients, the surgical team reported dramatic improvements in the symptoms of this horribly debilitating disease, with the beneficial effects lasting up to 3 months. As a result of this daring procedure and encouraging clinical outcomes, *The New England Journal of Medicine* published these two small papers based on limited understanding of the long-term effects, medical complications, and serious risk of death associated with this treatment. Virtually overnight, this work was reported by newspapers and television news programs around the world, leading to both increased interest from desperate patients and families, as well as from neurosurgeons who saw this new procedure as a novel palliative treatment for PD.

Within a few years time, many of these procedures were being carried out around the world, essentially unhindered by any well-designed controlled outcomes studies! Within 10 years, thousands of these operations had been completed, but the actual risk of morbidity and mortality remained unclear. In 1989, a surgical team from Chicago reported on a series of 19 patients who had undergone this procedure, and they reported that although some clinical improvement was observed, it was not as significant as that originally reported. Moreover, their patients could not cut back on their antiparkinsonian medications, the postoperative morbidity was high, and they cautioned against further widespread use of this procedure "outside of research centers." In a critical review of surgical therapies for PD, published in 2000 by the *Archives of Neurology*, Anthony Lang noted that the field had yet to produce a single published trial of surgery for PD "comparing patients randomized to surgery with those randomized to best medical management." By this time, the interest in adrenal medullary transplant surgery for PD, which was seen by surgeons and desperate patients as an exciting breakthrough, following the publication of only four cases in the most prestigious of all medical journals, had fallen out of favor – leaving a history of hundreds of surgical complications and deaths as a legacy. In this instance, both scientists and journalists were well intentioned in their wish to find relief for patients with a seriously debilitating disorder. But both rushed to publication without adequate scientific and journalistic restraint and in so doing, exposed many patients to risks of increased surgical mortality, morbidity, and falsely exaggerated hopes for cure.

Recent examples of troubling reports of research results in the media are not limited to only to the actions of scientists, surgeons, or journal editors. Moreover, at times the problem may stem not from what is published, but from what has been withheld from publication. Just recently, in 2004, a leading pharmaceutical corporation was sued for fraud by the Attorney General of the State of New York, with the claim that the company intentionally withheld (from the public view) data from large clinical trials showing that their highly profitable antidepressant medication is ineffective for the treatment of children and adolescents. Indeed, in some instances, the medication appeared to have negative therapeutic effects with an increase in suicidal ideation and behavior, data that were noted in early trials but not made available in public reporting. This case raises important ethical questions centering on the responsibility of publicly held corporations to release medical data that might alter the treatment practices of physicians, even if the release of such research results might negatively impact on the sales of their marketed drugs.

A third example stems from the general enthusiasm about new techniques for studying the functioning brain. Contemporary neuroscience research brings substantial media attention. Indeed, throughout the history of the field, depiction of advances in neuroscience in the media have influenced public attitudes toward biomedical innovations[12] and shaped ethical debates regarding the boundaries of neurological interventions.[13] A recent student examining web-based coverage of new technologies for studying brain function including positron-emission tomography (PET), neurostimulation, and neuropharmacology between 1995 and 2004 found that media reports in both the United States and United Kingdom were predominantly positive and enthusiastic about the findings and promise of these methods.[14] Through such reporting, the technical limitations of the imaging technologies are not publicly discussed and thus the public's ability to evaluate fully the perceived optimism of new technologies.[15,16]

Following the presentation of several contemporary case studies, we will turn our attention to one historical vignette that has been selected for its intrigue, complexity, and the nuance with which key issues were treated in the presentation of the study results. In 1964, José M.R. Delgado, a neurophysiologist at the Yale University School of Medicine, completed a demonstration experiment to prove to a skeptical public, and to the scientific community, that he could turn a brave, angry, and dangerous beast into a docile animal with the push of a button. On a bright spring afternoon, Delgado, both academician and performer, stepped into a Spanish bullring unarmed, so that a "brave bull" would charge at him with all of its "destructive fury." Delgado's demonstration of his ability to stop the bull before being impaled, by use of a hand-held transmitter that sent signals to electrodes buried deep within the animal's brain, was intended for a single purpose – to make headline news back home in the United States, and his effort proved to be quite successful. The story of Delgado facing the brave bull is a compelling vignette, which will frame a larger discussion on the complex relationship between researchers and both the electronic and print media. Delgado's story rose to public awareness as the era of psychosurgery (frontal lobotomies) was coming to a well-deserved close, and both scientists and clinicians were searching for effective next-generation interventions to treat dreadful psychiatric diseases. His pioneering work was widely hailed as groundbreaking, but the actual reports of his results with his brave bulls were incomplete and misleading. That said, his efforts were based on a true desire to ease the suffering of patients with serious mental illness and to control the aggression of dangerous criminals. Neither the scientist nor the journalists who reported on this story were ill-intentioned – each believed in the positive social force of (this particular) scientific discovery.

And finally, we have invited a panel of esteemed scholars from a variety of unique vantage points to consider these questions, and to offer their views on the interface between science and society, and how we might collectively strive to enhance and protect the public trust in science by ensuring that new discoveries are disseminated in a responsible manner. Our essayists provide fresh and creative perspectives, with diverse backgrounds that range from the front lines of the dispute between science and the Intelligent Design movement in the United States, to the boardroom of the world's largest pharmaceutical corporation.

SECTION

1

Historical and Contemporary Vignettes

John H. Warner, Ph.D.
Avalon Professor and Chair of the Section of History of Medicine
Yale University School of Medicine
New Haven, CT, USA

John Harley Warner is Avalon Professor and Chair of History of Medicine at the Yale University School of Medicine, and at Yale he is also Professor of History and of American Studies. He received his Ph.D. in History of Science from Harvard in 1984, and, after 2 years as a postdoctoral fellow at the Wellcome Institute for the History of Medicine in London, joined the Yale faculty in 1986. He teaches medical and undergraduate students and is a core faculty member of the Yale Graduate Program in the History of Science and Medicine. His books include *The Therapeutic Perspective: Medical Practice, Knowledge, and Identity in America, 1820–1885* (1986; 1997), *Against the Spirit of System: The French Impulse in Nineteenth-Century American Medicine* (1998; 2003), and the co-edited volumes *Major Problems in the History of American Medicine* (2001) and *Locating Medical History: The Stories and Their Meanings* (2004). Current projects include a study of medical student identity and dissection-room photographic portraiture, *ca.* 1880–1920, and a book on the transformation of the hospital patient chart in America, tentatively titled *Bedside Stories: Clinical Narrative and the Grounding of Modern Medicine*.

CHAPTER

2

Medicine, Media, and the Dramaturgy of Biomedical Research: Historical Perspectives

John H. Warner

1 Précis

By offering a historical perspective on relations among medical scientists, the mass media, and public perception and expectation, this chapter seeks to provide a broad context for understanding the publicizing of biomedical research as a cultural enterprise. While the conviction that medical science is progressive is centuries old, forged in the Renaissance, only since around the 1880s did the public widely come to expect biomedical breakthroughs – concrete advances that could make a real difference in their lives. By the early 20th century, the public expected to be spectators in an ongoing drama in which they held a stake, while medical researchers and media professionals in turn were assured that announcements of new discoveries would play to an audience primed to see these achievements as exciting and consequential. Focusing chiefly on the example of medical microbiology and on the United States, this chapter sketches in wide strokes changing patterns of media depictions of biomedical research. Stressing the co-development of modern medicine, modern media, and public faith in biomedical progress, it shows how such media as newspapers, magazines, health education pamphlets, radio, film, television, even comic books shaped and nurtured an image of research as a heroic drama punctuated by breakthrough discoveries. The aim is to recount and explain rather than to judge the simplifications and sometimes stylized myth-making that mediated between scientific practice and popular perception, in order to better understand how the researcher–media–public triad became an established feature of biomedical enterprise and the larger place of medicine in modern society.

It is June 2, 1881, in Pouilly-le-Fort, 40 km southeast of Paris. Crowds have gathered – entertained by acrobats and a brass band – around two pens of sheep inoculated with anthrax. Louis Pasteur and his family arrive by carriage. In one pen, all the sheep are dead. In the second, containing sheep that had received Pasteur's anthrax vaccine, the sheep appear dead until a dog barks, startling the slumbering flock to its feet to wild cheers from the crowd. Dr Joseph Lister warmly congratulates Pasteur. "Mad dog!" someone shouts, there are screams, and the crowd flees in terror.

This was a historic event that never happened. It comes from the Hollywood film *The Story of Louis Pasteur* (1936). Pasteur had indeed conducted a dramatic public demonstration of the anthrax vaccine in 1881, but what moviegoers saw was Pasteur as the film producers wished to remember him. The filmmakers could not have known that this famous episode was fictional – beyond artistic license – in a second sense: Pasteur had intentionally misled reporters, the public, and his colleagues. Having pinned his reputation to the promise of oxygen-attenuated vaccines, but yet to hit upon one that worked by the time of the public experiment at Pouilly-le-Fort, Pasteur resorted to a chemically treated vaccine very similar to the one publicly advocated by a competitor. By misrepresenting the nature of his vaccine, Pasteur won credit for the triumph over anthrax, diverted attention from the labors of his rival, and gained time to develop an effective oxygen-attenuated vaccine (which eventually he did). Only a historian's probing into Pasteur's private laboratory notebooks over a century later has revealed the deception.[1]

I want to suggest here a broad context for examining science–media relations by regarding the kind of depictions of research played out in 1881 and enacted on screen in 1936 as equally salient to understanding the constitution of public perception. Such media as newspapers, magazines, film, radio, television, even comic books have provided conduits through which reports about science selectively reached the public. But more than this alone, media representations of the process and products of research have informed popular expectations in ways that have been important in structuring media professionals' and scientists' assumptions about their audience.[2–4] Media publicizing of biomedical research, by informing public perception and expectation, was a key engine in the remarkable rise in prestige of medicine through the mid-20th century, just as inflated public expectations also nurtured by the mass media were one source of the disillusionment, criticism, and declining prestige that marked the final third of the 20th century. Understanding these patterns of assumption and expectation must grow from recognizing the co-development of modern medicine, modern media, and public faith in biomedical progress.

With the rise of mass media in the late 19th century, it became easy to find words cautioning against the seductive dangers of the media. In 1904 William Osler counseled his fellow physicians to avoid "the temptation to toy with the Delilah of the press... . There are times when she may be courted with satisfaction, but beware! Sooner or later she is sure to play the harlot, and has left many a man shorn of his strength, viz., the confidence of his professional brethren."[5] In his book on how to succeed in the business of medicine, D.W. Cathell (1916) more emphatically admonished his readers to "*steer clear* of editors, reporters, and interviewers; [and] determinedly *keep your name out of the newspapers*."[6]

Such wariness notwithstanding, doctors soon recognized that the mass media was becoming a critically important ingredient in the unprecedented cultural success of medicine. The media shaped and sharpened public expectations – deeply rooted assumptions about the nature of change in biomedical knowledge and its consequences. While the conviction that medical science is progressive was forged in the Renaissance, only since around the 1880s did the public widely come to expect breakthroughs. The public was newly enlisted as spectators to a drama that gained in meaning from the possibility that research might yield products that could make a real difference in their own lives.

The case of Pasteur is not chosen at random. Medical microbiology, my leading focus here, was the platform on which collaborations between scientists and media professionals in the late 19th century first forged the notion of biomedical breakthroughs and infused this idea into popular thinking. During the early decades of the 20th century, moreover, it was proselytizing for what historian Nancy Tomes has called "the gospel of germs" that conveyed to the public the most culturally powerful images of medical research, researchers, and their products.[7] I want to recount and explain rather than to judge the simplifications and sometimes stylized myth-making that mediated between scientific practice and popular perception. Looking chiefly at the United States, I will sketch in broad strokes the changing patterns of media depictions of biomedical science that established the context in which scientists, media professionals, and citizens all made choices, and within which those choices must be understood.

2 The germs race and the advent of biomedical breakthroughs

The emergence of the notion cultivated in the mass media and widely shared by the public that medical research would bring *dramatic* discoveries must be understood against the backdrop of late-19th century medical microbiology. The experimental methods that enabled Robert Koch to identify the anthrax bacillus in 1876 and the tuberculosis bacillus in 1882 initiated the discovery in rapid succession – at the rate of one a year between 1879 and 1900 – of the causative agents of such diseases as leprosy, cholera, diphtheria, typhoid fever, tetanus, plague, botulism, and dysentery. These discoveries did not impel a lasting transformation in public attitudes toward research, but rather provided the stage on which new relationships among medical scientists, media, and the public initially were played out.

Competition for priority not so much among researchers as among nations – most starkly the bitter rivalry between Koch in Prussia and Pasteur in France – drew media attention. Nationalism gave special incentive to be first in this germs race, and exerted considerable pressure to proclaim discoveries with haste and publicity. As historian Andrew Mendelsohn has argued, like Pasteur's push after the Franco-Prussian War to make a superior French beer (a "bière de la revanche nationale"), in the face of German bacteriological achievements, the Institute Pasteur was opened in 1888 as an "institute de la revanche."[8] Prussia returned the volley in 1891 by establishing the Institute for Infectious Diseases for Koch. "One can be an anarchist, a communist, or a nihilist, but not an anti-Pastorian," one lone French clinician critic of Pasteur complained in 1887; "a simple question of science has been made into a question of patriotism."[1] Sectional rivalries could also drive a sense of urgency in announcing etiological discoveries and muster eager press attention. For researchers in the post-Civil War American South, for example, where yellow fever threatened life and commerce, the hunt for the yellow fever germ was closely bound up with regional pride. During the 1880s and 1890s, before the viral etiology of the disease was known, the bacterium that causes yellow fever was discovered, debated, and abandoned more than a dozen times.[9]

Pasteur's 1885 announcement of having successfully treated dog bite victims at risk of rabies marked a watershed in popular expectations of biomedical change. His earlier successes in developing vaccines against such animal diseases as chicken cholera, anthrax, and swine fever had sparked the hope that vaccines for human diseases might be found as well, and his rabies cure seemed a fulfillment of this dream. "From the outset, Pasteur knew that he would be hailed as a savior if he succeeded in this quest," historian Gerald Geison has observed. "Here above all he displayed the theatrical flair that marked his choice of subject to pursue and his manner of presenting the results to an audience gripped with suspense and eager to hear a happy ending."[1] Rabies was a rare disease, killing less numbers of people compared with such killers as tuberculosis, but its horrifying symptoms and uniformly fatal outcome inspired in the public a terrible fascination. It was the nationalistic fervor his rabies vaccine incited that led to the creation of the Institute Pasteur, and donations to support his rabies work poured in. At his death, Pasteur would be mourned at a state funeral as a hero of France.

In the United States, newspapers reported Pasteur's initial announcement, but with short notices and little fanfare. Then in December 1885, when four working-class children in Newark, New Jersey, were bitten by a "mad dog," Pasteur's rabies treatment suddenly became front-page news across America. A local physician urged that the boys be sent to Pasteur for the only treatment might save their lives, and within days they were aboard a steamer for Europe. The press orchestrated public fund-raising and the boys' adventures were lavishly covered by newspapers across the country, with pictures of the new celebrities, as well as by heavily illustrated magazines. An exhibit at a popular New York wax museum commemorated their miraculous cure while they were still abroad, and the boys went on to tell their own story onstage to paying crowds at working-class "dime museums" in New York and Philadelphia. The media elevated the episode to a cause célèbre, and kept the story alive for months after the boys' healthy return through news articles, features, editorials, illustrations, letters to the editor, jokes, cartoons, songs, and even political satire.[10,11]

This media spectacle arose from a convergence of developments in medical science and in the mass media. New technologies from mid-century, particularly the steam-powered high speed rotary presses, were joined by the telegraph and laying of a transatlantic cable, which facilitated mass circulation of fresh stories. So too, the shift from the recently developed human-interest story into unabashed sensationalism, emergence of editorial novelties such as fund-raising subscriptions, and new printing techniques that made it easy to publish pictures – all innovations enlisted in telling and retelling the rabies-cure drama – helped transform a report about medical research into a widely visible media sensation that could capture the popular imagination. The rabies drama "changed the thinking of newsmen, medical leaders, and the general public alike," historian Bert Hansen has argued, "and firmly planted in mass culture expectations about the nature and meaning of a medical breakthrough."[11] The patterns of expectation that emerged transcended Pasteur and his work, creating new, widely shared assumptions about the benefits medical research would bring and setting the template for medical and media management of subsequent discoveries, such as diphtheria antitoxin (1894) and X-ray (1896).

Celebrated medical breakthroughs did not always prove to be right. In the fall of 1890, for example, Koch, pressured by the Prussian Kultusminister, announced the discovery of what he called tuberculin as the first effective remedy for tuberculosis.[12] This cure for the leading killer of the age drew riveted attention from newspapermen, physicians, tuberculosis sufferers, and the wider public. A frenzy of media publicity ensued, in Europe and in America, where Koch's announcement – cabled from Berlin – appeared in translation on the front page of the *New York Times*.[11] Just as the French state had created an institute for Pasteur in the wake of his rabies vaccine discovery, so Prussia made plans for creating a Royal Institute for Infectious Diseases for Koch. Pasteur, the consummate entrepreneur, capitalized on French media attention to the Prussian achievement to initiate plans to expand his own institute. Within a few months, though, tuberculin proved to be a clinical failure as a therapy (although useful as a diagnostic tool), bringing an angry backlash against Koch. The momentum for establishing his institute had gone too far to be halted and the Berlin Institute opened in 1891. Pasteur, however, had to cancel his expansion plans, complaining that "the failure of Koch's remedy must put an end to our project."[8] Yet, while the tuberculin affair ended as a fiasco, it displays how widely medical, media, and public expectations were shared by 1890, so much so that they clearly played a role in pressuring Koch into what in retrospect was a premature announcement.

By the start of the 20th century, then, expectations that scientific research would lead to medical progress, and that the accomplishments of medical science would be heralded by the media and celebrated by the public, were anchored in place. Certainly that was clear in 1900, when, in an achievement assertively claimed as an American breakthrough, the Yellow Fever Commission led by Walter Reed announced on the front page of the *New York Times* its conclusions about the role of the *Ades aegypti* mosquito in transmitting yellow fever. Breakthroughs were quintessentially social phenomena, engaging researchers, media professionals, and public. "The public's new expectations of medical miracles engineered at laboratory benches," Hansen has asserted, "came to shape the actions of scientists, physicians, and journalists."[10]

3 Fashioning research heroes

Starting in the 1880s and 1890s, American medicine experienced an enormous elevation in public esteem and cultural authority. Historians recently have eschewed earlier, deterministic accounts that attributed this uplift to greater efficacy. Despite some notable exceptions (such as diphtheria antitoxin in 1894 and Salvarsan in 1909), elevation in standing came *before* the new experimental sciences had delivered the goods in terms of greater power to heal. By and large, hopes for effective antimicrobial therapies remained unrealized until the development of sulfa drugs and antibiotics in the 1930s and 1940s. The question, accordingly, becomes: What accounts for the remarkable rise in the fortunes of American medicine?[13]

While the answers are complex, the new relationship between medicine, media, and the public was an important cultural leaven. Not only the underlying public expectation of progress but also the shared assumptions that the results of scientific research should be *news* on a regular, continuing basis were engines for change. By the early 20th century, the public expected to be spectators in an ongoing drama in which they held a stake, while medical researchers and media professionals in turn could expect that their announcements of new discoveries would play to an audience primed to see these achievements as exciting and consequential. The researcher–media–public triad had become an established feature of the biomedical enterprise and the place of medicine in American society.

The work of biomedical scientists was most vigorously and visibly conveyed to the public not by individual reports on putative advances, but through the mass health education campaigns that became a prominent feature of American life.[14,7] During the early decades of the 20th century, health messages were circulated by the full range of print media, which enlisted the new visual culture techniques developed for advertising, and by roadside billboards, displays at county fairs, and placards posted on streetcars. Messages were taken to the streets in health parades and children carried and received personal and public hygiene lessons through health plays, rallies against the fly, and demonstrations of healthy living habits. Health lessons were also urged through the new medium of public entertainment and education, motion pictures, starting with a series of films that in 1910 the National Association for the Study and Prevention of Tuberculosis commissioned Thomas Edison to produce, such as *Hope* in 1912.[15] As part of the wider program for selling germ theory and its implications to the public, anti-tuberculosis societies launched the first mass health education campaign in American history. By 1922 the Modern Health Crusade, inaugurated in the mid-1910s as a children's crusade against tuberculosis that featured plays, games, and songs, had enrolled over 7 million American children. "Children," one Brooklyn campaigner commented, "are a very good advertising medium."[7] Modern publicity methods, a sensibility to spectacle, and the sensationalism characteristic of the emergent age of advertising all informed the campaign to infuse the lessons of medical microbiology and hygienic living into the workaday routines of ordinary Americans.

It was here more than at any other site that biomedical research, the modern mass media, and the techniques of the advertising industry that accompanied the rise of consumer culture came together.[16] The confluence of these impulses gave medical investigators unprecedented visibility in American society. More than this, while the aim of these health education campaigns was to mold behavior, in the process they further reinforced and shaped public expectation of progressive change and continued breakthroughs.[17]

One device used in these campaigns that became commonplace in the 1920s was the portrayal of medical scientists as heroes. The full array of mass culture media – from cheap print to film – was deployed in glorifying the heroes of medical science as vehicles for health education and as role models. The Metropolitan Life Insurance Company began to publish a widely circulated series of pamphlets called *Health Heroes*, each devoted to the life of an exemplary scientist – Pasteur, Koch, Reed, and Marie Curie among them. Selectively and dramatically retelling a story of the obstacles they overcame, their discoveries, and their enduring contributions, these renderings of scientific heroism equated medical progress with breakthrough discoveries, linked health to research, and attached the authority of practicing physicians to medical science.[7] The series was still circulating after World War II, joined by short companion film strips for classroom use. Mirroring wider trends in science popularization – especially the shift away from science as process and method toward science as event and as product that historian John Burnham has underscored – both researcher-heroes and their discoveries became media commodities.[18] Maneuvers such as these made it all the easier for media professionals to present medical scientists in a heroic mode, assured of finding a receptive popular audience, while at the same time this widely disseminated portrait of research as a heroic drama punctuated by breakthrough discoveries invited scientists themselves to adopt the role that the media, the public, and their predecessors had scripted for them.

4 Microbe hunting in Hollywood

The publicizing of scientific research in newspapers and popular magazines experienced an enormous upsurge in quality and quantity of reporting starting early in the 1920s. This change was associated with the appearance of science reporters, a new breed of journalists typically known in the office as "Doc," and of a special science news service.[18] Newspaper publisher W.W. Scripps had become convinced that most Americans ("the 95 per cent," he called them) received the bulk of their education not from schools but from the daily newspaper. Science coverage in the press, however, was slim and full of "misinformation," and "the tales of [scientists'] adventures, dramatic as they are, seldom find their way into print." Science Service, founded with his support in 1920, was to be an interpreter. Playing to popular interest but reacting against the press "ballyhoo," Science Service cultivated a style of writing that, as historian David Rhees has put it, "replaced sensationalism with the romance of facts, and science was transformed from lurid side-show to high adventure."[19] The Director of the Service began in 1929 to circulate to editors and correspondents an annual list of "Stories to Be Careful Of," including telepathy, cancer "cures," and man-eating trees. By 1934, the community was sufficiently large and professionalized to create the National Association of Science Writers.

No science journalist of 1920s and 1930s did more than Paul de Kruif to instill in the popular imagination an image of biomedical research as drama, romance, and adventure. De Kruif held a PhD in bacteriology, conducted research at the Pasteur Institute during World War I, and then entered a research career at the Rockefeller Institute.[20] His early efforts at journalism – which attacked the pretensions of American physicians, even while celebrating the ideals of scientific research – brought him into conflict with the head of the Rockefeller Institute, Simon Flexner, and in 1922 de Kruif left laboratory science to become a full-time writer. Expanding on magazine articles, in 1926 he published *Microbe Hunters*, a collection of a dozen dramatizations of bacteriological discovery focused on such investigators as Pasteur, Koch, Reed, and Paul Ehrlich. "His scientists are adventurers, explorers, heroes," a reviewer in the *New York Times Book Review* rightly observed, and the book (in print ever since), like the overarching portrayal of research as drama, was phenomenally successful.[21] "You've hit the jackpot, my boy," H. L. Menken told him privately; "You're all alone in a new field."[22] Yet, the field quickly became populated, and for the next couple of decades stories by popular writers of biomedical research as a romantic and heroic enterprise filled the stands of American bookstores.

Even before de Kruif began writing *Microbe Hunters*, his collaboration with novelist Sinclair Lewis had enhanced the image of scientist as hero by introducing the fictional but widely influential heroic protagonist of *Arrowsmith* (1925). Martin Arrowsmith is a biomedical researcher, the first of consequence in American fiction. De Kruif, in a consulting role, brought to the Pulitzer Prize winning novel both technical advice on science and a roster of real-life prototypes for the book's characters. Above all, though, he brought his reverence for pure research as a moral enterprise, a redemptive counterpart to the materialistic temptations and corruptions that permeated American life, including medicine.[23,24] The reviewer for *Science* celebrated the novel as an important and welcome indicator of "a certain shift in our civilization shown by the growing interest of the layman in scientific matters," noting that "with the coming of this interest, suspicion has given way to support."[25] More than this, the novel reinforced a growing public perception of biomedical research as a heroic endeavor, and of the true scientist as embodying not only technical expertise but also virtue.

The popular profile of medical research as heroic drama received an enormous cultural boost when, in 1931, *Arrowsmith* opened as a full-length feature film. Directed by John Ford, starring Ronald Coleman, featuring Helen Hayes, and introducing Myrna Loy, the screenplay, adapted by playwright Sidney Coe Howard for United Artists, simplified the novel but retained the core of Lewis's story: the medical scientist's spiritual struggle and moral achievement.[26] By the early 1920s, medical professionals had clearly recognized the power of film as a medium to influence public attitudes, and the box-office success of *Arrowsmith* inaugurated the heyday of American medicine's relationship with Hollywood.[27]

The popularity of *Microbe Hunters* and *Arrowsmith* inspired a stream of full-length feature film biographies – known as "biopics" – of past medical researchers; among them are *The Story of Louis Pasteur* 1936, *Yellow Jack* – on Reed and yellow fever (1938), and *Dr. Ehrlich's Magic Bullet* (1940). *Pasteur*, the first, garnered three Academy Awards for Warner Brothers, including Best Actor for Paul Muni in the title role, and was an enormous money maker. What all of these films shared was a depiction of the protagonist's hard work, perseverance in the face of skeptics and other adversity, and virtue (rewarded in the end), along with an enduring breakthrough that benefited humanity.[27,28] This was true as well of the RKO film *Sister Kenny* (1946). Australian nurse Elizabeth Kenny, however, whose heterodox polio treatment gave her in real-life public stature as a discoverer struggling for acceptance just as it won for her ranking in Gallup Polls as most admired woman in America, above Eleanor Roosevelt, was not a historical figure but alive and actively involved with scripts, filming, and even the hand movements of Rosiland Russell, who played her in the starring role.[29,30]

The image of medical researchers cultivated in Hollywood films gained in power from reciprocal reinforcement with other media, including radio, which grew important in the 1930s. "This blurring of genre contributed to greater integration of both characters and images into popular culture," historians Susan Lederer and Naomi Rogers have noted.[27] "Both medical and media professionals (scriptwriters, producers, and reporters) participated in dissolving the boundaries between entertainment and education." Drawing from the film *Arrowsmith*, actors Spencer Tracy and Fay Wray portrayed the main characters on the Lux Radio Theater, which would go on to bring a long string of heroic discoverers – Pasteur, Reed, Ehrlich, and Kenny among them – into American living rooms.[26] Howard, who wrote the screenplay for *Arrowsmith*, had in 1927 begun preliminary notes for a stage play to be titled *Yellow Jack* (inspired by de Kruif's treatment of Reed in *Microbe Hunters*), which opened on Broadway in 1934. When in 1937 Metro-Goldwyn-Mayer purchased the film rights, Howard was unavailable only because he was busy adapting the novel *Gone with the Wind* for the screen.[27] And after *Dr. Ehrlich's Magic Bullet* (with Edward G. Robinson in the title role) was released in 1940 – centered on his discovery of Salvarsan as a treatment for syphilis – the US Public Health Service made a short, half-hour version of the Warner Brother's film as part of its World War II national campaign against syphilis.[28]

"The dramatization of medicine on the stage and screen was a response to popular tastes already apparent," historian Richard H. Shryock observed.[31] The success of medical-themed films like *Men in White* (1934), *Women in White*, and *The White Angel* (1936) made it seem "as though anything 'in white' was good for box-office returns." A whole generation was recruited to microbiology and medicine by reading *Arrowsmith* and *Microbe Hunters* as teenagers. More than this, media portrayals both profoundly shaped public understandings of how science works and structured relationships between scientists and the public.[32]

5 Great moments

Media depictions of medical research helped sustain what historians have called "the Golden Age" of American medicine. This was the era in the middle decades of the 20th century when medical institutions and practitioners enjoyed their greatest esteem in American society – public confidence and cultural authority surpassing anything experienced before, or since. And once wonder drugs did begin to appear – sulfa drugs in the 1930s, antibiotics in the 1940s – they were taken as but confirmation that faith had been well placed. Public trust seemed boundless, owing in no small measure to the ways the media had publicized biomedical research and nurtured high expectations of continued progress.[33–35]

Media treatment of medicine remained overwhelmingly positive, seldom overtly critical, and in postwar America the celebration of biomedical researchers and their deeds intensified. When in the 1940s comic books rose to mass popularity, playing a major role in the imaginative world of American children and young adults, one current in the industry published on "true adventures," dramatizing the exploits not of fantasy superheroes but real people. Starting with a 1941 *True Comics* that featured Reed, the stories of figures like Ehrlich, Koch, Kenny, and especially Pasteur turned upon a quest for discoveries portrayed as a valiant form of heroism.[36] Drug companies also cultivated these images of biomedical research. In 1951, for example, as part of public relations and advertising campaigns, Park, Davis and Company commissioned a series of oil paintings portraying what it called "Great Moments in Medicine," and during the 1950s and 1960s laypeople encountered mass produced and widely circulated copies commemorating the achievements of figures like Pasteur, Reed, and Ehrlich on the walls of doctors' waiting rooms across North America.[37] The intense popularity of the new medium of television in the 1950s, with the appearance of prime time medical dramas (though featuring practicing doctors more than researchers), reinforced a positive image of medicine. Organized medicine had long worked closely with media professionals in crafting and censoring Hollywood films, and the rapid spread of television prompted further moves to standardize and regulate media. In 1955 the AMA added television to its Physicians' Advisory Committee on Television, Radio and Motion Pictures, which reviewed scripts and script rewrites to assure that the medical images and behaviors that reached American living rooms were positive.[27,38] Public opinion polls ranked physicians as among the most highly admired professionals in America, placing them equal to or higher than Supreme Court justices.[34]

This was the context of popular expectation the media had fostered when, in the mid-1950s, the public heard about the Salk vaccine against poliomyelitis. Public eagerness for the vaccine made it difficult for scientists who wanted to proceed cautiously, especially in a Cold War climate in which this research breakthrough by a scientist-hero seemed an affirmation of American scientific progress. The *New York Times Magazine* called the Salk vaccine trials the "climax of a stirring medical drama." As historian Allan Brandt has observed, "no medical experiment ever held such public attention," and a Gallup poll revealed that more Americans knew of the field trials than could identify the full name the President of the United States.[39] It was a media event, one that both drew upon and sustained sensationalism in reporting on biomedical research.

By 1966, one observer could quip that "few newspapermen are interested in a scientific finding unless it can cure cancer while in orbit."[18] When in 1964 Yale neurophysiologist José M.R. Delgado stepped into a bullring to demonstrate how he could stop a charging bull by pressing a button – sending a signal from a hand-held transmitter to an electrode planted in the animal's brain – if his media stunt seemed a performance scripted by Hollywood, there were sturdy cultural reasons for the resemblance: he drew upon tropes (ranging from breakthrough as event to scientist as hero) that had grown familiar to scientists, media professionals, and the public (please see Chapter 3, for a thorough review of this remarkable event). Yet, even as his photograph appeared on the front page of the *New York Times*, signs of a new critical skepticism about the trust placed in medicine by the American public were becoming unmistakable. By the 1970s and 1980s, many observers pointed to a crisis of confidence in the medical establishment and its cultural authority.

Sensationalist reporting had helped to fuel anti-medical movements by raising public expectations to unrealistic levels. Of the medical discoveries reported in the media, according to one mid-1960s study, 90 per cent ultimately failed.[18] By the time *Health* magazine announced in 1983 its "Fourth Annual All-Breakthroughs Issue," expectations of breakthroughs from medical research first forged a century earlier had reached a level calculated to disappoint.[18] Portrayals in the mass media not only reflected but were constitutive of declining stature, with negative depictions of medicine becoming common. Medical scientists' concerns about the publicizing of biomedical research may also have been a vehicle for expressing wider anxieties about a public increasingly ambivalent about the enterprise they once revered as heroic.

Pasteur had intentionally misled the press, the public, and his peers, not only in announcing his anthrax vaccine but also in launching the rabies vaccine that set the model for biomedical breakthrough and became the leading cornerstone of his deserved fame. Historical study of his laboratory notebooks has only recently revealed that the successful vaccine trials he claimed to have performed on a large number of rabid dogs never took place, and that Pasteur had never tried on animals the vaccine used in treating Joseph Meister, the renowned and successful first trail of rabies vaccine on a human. Moreover, by the time Meister was treated, Pasteur had already used another rabies vaccine on two other human subjects, one of whom died, but he never mentioned these experiments publicly. His vaccine ended up working. But one might even conclude that part of Pasteur's *success* as a biomedical scientist hinged upon his strategic engagement in "intentionally inexact reporting," if not outright lying.[1]

Sorting out and judging the ethics of Pasteur's "private science," however, is less important for researchers today than recognizing the broader patterns of interaction over the ensuing century among medical researchers, the media, and the public that his work helped instigate. The media has been much more than either a helpful tool for scientists to use or, conversely, a source of potential problems to be managed. By shaping public perception and expectation, the mass media has played a key and active role in the cultural success of biomedicine. Understanding the workings of the medicine–media–public triad in the past offers no clear chart for the future, but does begin to suggest the extent to which the place of biomedical research in society will continue to be grounded in a dynamic relationship among medicine, media, and popular expectation.

CHAPTER

3

Delgado's Brave Bulls: The Marketing of a Seductive Idea and a Lesson for Contemporary Biomedical Research[a]

Peter J. Snyder

> *Scientists have been learning to tinker with the brains of animals and men and to manipulate their thoughts and behavior. Though their methods are still crude and not always predictable, there can be little doubt that the next few years will bring a frightening array of refined techniques for making human beings act according to the will of the psychotechnologist.*
>
> Boyce Rensberger
>
> *The New York Times*, September 12, 1971, Sect. E, p. 9

With the fall of the cortical equipotentiality hypothesis, championed by Karl Lashley[1] and many of his colleagues in the 1920s and 1930s, and commensurate with the re-acceptance of the cortical localization models that have

[a]I wish to express my gratitude to Dr. Toby A. Appel, the John R. Bumstead Librarian for Medical History, at the Harvey Cushing/John Hay Whitney Medical Library of Yale University (New Haven, Connecticut), for her assistance in obtaining articles, letters, and press releases detailing Prof. Delgado's career at the Yale Medical School. I also deeply appreciate the wise advice, support, and encouragement of Drs. Harry A. Whitaker, Lauren Julius Harris, Sidney Segalowitz, and Elliot Valenstein throughout this project. This essay was originally prepared as a symposium presentation, for the 12th annual meeting of TENNET (Theoretical and Experimental Neuropsychology/ Neuropsychologie Expérimentale et Théorique), Montréal, June 2001.

been shaped and refined over the past seven decades, neuroscientists have sought to develop methods for the stimulation of discrete neural structures in order to both modify and control behavior and to treat psychiatric illness. With the exception of pioneering efforts by J.R. Ewald toward the end of the 19th century, aimed at stimulating the brain of awake and behaving dogs with thin intracerebral wire electrodes,[2] the technique of placing electrodes in the brain to study or modify behavior was little noticed until the early 1930s, when the Swiss physiologist Walter R. Hess reported on his method to explore the hypothalamus and other structures in unanesthetized cats.[3] In his experiments, Hess demonstrated that autonomic functions, movement, sleep, and (putatively) both fear and aggression might be influenced by the electrical stimulation of specific cerebral structures.

Hess's publications on this topic immediately preceded the landmark work by Frederick Gibbs[4] who reported that, by passing a small current through electrodes placed in the hypothalamus (probably located within the lateral portions of this structure and what later became known as the "pleasure center"), he was able to elicit "purring" reactions in cats.[5] This line of investigation then appeared to fall dormant for approximately two decades, until the rapid expansion of the fields of psychosurgery, psychopharmacology, and physiological psychology in the early 1950s.

In 1954, James Olds and Peter Milner reported on their fortuitous discovery, at the Montreal Neurological Institute, of positive and self-reinforcing electrical stimulation of the septal region of the rodent brain using a self-stimulation test paradigm.[6] This largely accidental discovery by Olds and Milner in 1953 (at the time, Olds was serving as a post-doctoral fellow in the laboratory of Donald Hebb), has been well-reviewed elsewhere.[7] These young investigators were implanting electrodes into the reticular activating system (RAS) of rats, in order to explore the contributions of the RAS to spatial learning by increasing autonomic arousal. As part of their investigation, they stimulated the animals to see whether they could replicate reports by Prof. Neal Miller and a young physiologist with whom he was collaborating, José M.R. Delgado (both from the Yale University School of Medicine) of the elicitation of a fear response. Miller and his colleagues had already shown that electrical stimulation to portions of the reticular formation led to aversive and negative-reinforcing responses.[7] In a now famous and fortunate mishap, one electrode was unintentionally bent as it was being directed toward the RAS of one particular rat, with the electrode tip apparently placed in or near the hypothalamus. Unfortunately, this specific brain was apparently lost and histological verification of the actual electrode placement was never completed. At any rate, Olds and Milner stimulated the electrode in this one rat and essentially observed the opposite of the expected fear response described by Miller, that is, they discovered and then verified the existence of the "pleasure center" in the brain. They later confirmed this initial finding with subsequent experiments utilizing the self-stimulation paradigm with many more rodents.[8] As we now know, the role of the septal area, an evolutionarily old structure, in pleasure and reward, has been replicated in species ranging from chickens and rabbits, to dolphins and humans. Furthermore, this brain region, rich in dopaminergic and glutaminergic activity, has been repeatedly implicated in the genesis of addiction in both humans and animal models.[b]

In the early 1950s, Delgado sought to collaborate with the learning theorist and psychologist, Neal Miller, at Yale, to show that they could elicit a "fear response" with electrical stimulation of the brain (ESB) in cats, most likely by stimulating portions of the hypothalamus and limbic system close to the midline of the brain.[8] Delgado's objective, in this work, was to identify a common underlying circuitry for negative reinforcement that is activated by any aversive stimuli. This work was enormously exciting, and captivated the imagination of both the medical and lay communities, with a resulting explosion in research on methods to control behavior and treat psychiatric illness. These early discoveries could not possibly have been better timed, as by the mid-1950s the era of psychosurgery and the frontal leukotomy was in full swing. By 1950, Drs. Walter Freeman and James Watts, of George Washington University in Washington, DC, had operated on over 1 000 patients, with the rate of these surgeries steadily increasing over the next 5 years. By the late 1950s, when psychosurgery began to be seriously questioned and its use tempered by the advent of effective psychopharmacologic agents and the increasing reports of iatrogenically based adverse sequelae, well over 40 000 patients had been subjected to frontal leukotomy surgeries in the United States alone.[8] Clearly, there was great interest in the development of alternative therapies that were less invasive and not associated with the adverse side effects of the surgical procedure, and ESB – with one of its most vocal champions being José M.R. Delgado – was touted as an excellent alternative.

1 José M.R. Delgado

Delgado was born in 1915 in Ronda, Spain, he completed his medical degree at University of Madrid in the 1930s, and he remained there on

[b]As an interesting aside, although the discovery of the pleasure center is credited to Olds and Milner, it appears to have been made initially by the psychiatrist Robert G. Heath and his colleagues at Tulane University. Between 1950 and 1952, Heath's group implanted electrodes into the brains of 24 schizophrenic patients. For four of these patients, stimulation of specific electrode leads led to descriptions of the experience as "pleasant" or leading to feelings of being "jovial" or "euphoric". For all four of these cases, the area that was stimulated was the septal region. Although Heath reported this as a brief note, contained in a 1954 publication,[9] he took little notice of this finding as he likely perceived this to be irrelevant to the specific neuropathological model of schizophrenia that he was testing with these studies.[7]

faculty (in the Department of Physiology) until 1950. In 1946 he accepted a Fellowship position at Yale University in order to work with the physiologist Dr. John Fulton. It was John Fulton, along with Carlyle Jacobsen, who showed that bilateral frontal lobectomy surgery dramatically alters aggressiveness in the chimpanzee, despite the fact that the original intent of their research was to study the effects of frontal lobe resection surgery on learning and short-term memory. In fact, it was Fulton's original paper presentation at the 2nd International Congress of Neurology (London, 1935) that convinced one very interested member of their audience, Dr. Egas Moniz (1874–1955), to first attempt frontal lobe surgery in human psychiatric patients.[10]

In 1950 Delgado was offered a regular faculty appointment at Yale, in the Department of Physiology headed by Fulton. Beginning in the early 1950s and extending over several decades, Delgado's laboratory pioneered techniques for both electrical and chemical stimulation of the brain (across a range of animal species, including humans), with the aim of studying and manipulating arousal, consciousness, affect, basic instinctual drives, and mood states. From this early stage in his career, Delgado worked to distinguish himself as a leading investigator in this burgeoning field of study, and by 1952 he had earned Spain's most distinguished biomedical research honor, the Ramón y Cajal Award. Moreover, the relatively long list of newspaper articles and press releases from Yale University in the early 1950s attest to what became his well-known interest in seeking media attention for his work.

Early in his career, Delgado saw a practical application of his work in the development of ESB methods for controlling satiety and caloric intake. *The New York Herald Tribune* reported on his early work on this topic, in collaboration with Prof. C.N.H. Long (also from Yale University), with the title of the article being "Overeat? Blame the Hypothalamus."[11] By March of 1954, Delgado had started to receive national television coverage for his research, with the *New Haven Evening Register* publishing an article with the title announcing, "Yale Medical Laboratory to Take Part in Nationwide TV Diet Demonstration: Dr. Delgado, Using Monkeys, Will Show the Effect of Brain on Eating Behavior."

Delgado's penchant for media attention was well known amongst his colleagues at the medical school, and his style was at times viewed as controversial. Delgado's primary academic appointment was in the Department of Physiology, where he became an Assistant Professor in 1953, and rose to the rank of Full Professor in 1966, a position that he remained in until he left Yale to return to the University of Madrid in 1974.[12,13] He also held a joint academic appointment in the Department of Psychiatry at Yale, although his relations with the faculty in that department were strained at times. As one of his former colleagues, an esteemed scientist who occupied an office very close to Delgado's, put it discreetly to me, "He kept his lab space and support from the university, in part, because of whom he married." In fact, in the early 1950s Delgado married a lovely young woman, Caroline Stoddard, who would serve as a Research Assistant in his laboratory for many years afterwards. Stoddard's father, Mr. Carlos F. Stoddard, initially served in the University's development office, but by 1964 he rose to become the Director of Endowment and Gifts for Yale University, a powerful position reporting to the President of the university. These comments are not meant to diminish, in any way, the important scientific contributions that Prof. Delgado made over the course of his long career. Rather, these observations, which have been privately echoed by several of his prior colleagues at Yale, are offered in attempt to place into context his decision to stage the theatrical public demonstration that he orchestrated, in 1963, on a bull ranch in southern Spain.

Regardless, Delgado was a brilliant and creative scientist, and his research program grew rapidly at Yale. He required a large amount of space, including a large monkey colony, two operating rooms, space for several laboratories, and funding for a relatively large cadre of students, research assistants, and post-doctoral fellows. Fortunately for him, he chose to focus on an area of study that almost instantly captured the collective imagination of the media, the public, and government agencies – including two agencies within the Department of Defense that funded his research. Very early in his research program, Delgado began to collaborate with others who shared his interest in determining whether his ESB techniques could be useful in diminishing the frequency and severity of positive symptoms of psychosis, and to encourage prosocial behavior in patients with severe psychiatric illnesses, such as individuals prone to violent acts of aggression.[14–16]

On April 09, 1953, at a meeting of the American Physiological Society held in Chicago, Illinois, Delgado delivered an oral presentation describing his new methods for implanting between 7 and 28 electrodes in various sections of the primate brain (typically rhesus macaques). A press announcement, released by Yale on the same afternoon, boasted of a "new technique of implanting electrodes in the brain [that] gives promise of opening up some of the mysteries surrounding the function of the brain in man and animals."[17] By the evening of that same day, the *New Haven Evening Register* published an almost verbatim transcript of Yale's press release,[18] and 3 days later an account of Delgado's presentation and claims was published by *The New York Times*.[19] A day later, on April 13, 1953, an article, entitled "The Ocean of the Mind," was published in *Time Magazine* hailing the groundbreaking research of J.M.R. Delgado, and noting, amongst other tidbits of information, that, "The seat of a monkey's love for bananas evidently is deep in the frontal lobe of the brain: current applied here will make him refuse bananas."[20] The author of this Time Magazine article reported that when Delgado stimulated the hippocampus of a "ferocious" macaque, the animal was turned "into a macaque Milquetoast; he even let Dr. Delgado take the liberty of stroking his face. The moment the current was turned off, he tried to bite." The article concluded by presenting the obvious and exciting ramifications of this work for the treatment of severe psychiatric illness, and by describing some of the largely uncontrolled clinical experiments that were being conducted by other researchers concurrently, on human patients with schizophrenia and epilepsy (e.g., Dr. R.G. Heath at Tulane University [see footnote b], and Dr. R.G. Bickford at the Mayo Clinic in Rochester, Minnesota). Hence, within a matter of days, Delgado was catapulted to the national spotlight for perhaps the first time in his career as an established and independent investigator – a position that he would continue to cultivate until his departure from Yale Medical School in 1974.[13]

Between 1954 and 1964, Prof. Delgado's laboratory work and fame continued to flourish, with a profusion of published reports detailing his successes in modifying the social, eating, sleeping, and "emotional" behaviors of freely behaving animals, with a particular focus on monkeys. A number of his more important papers are reviewed in his published address to the New York Museum of Natural History, in 1965,[21] and are also reviewed in exhaustive detail in his now famous book, entitled, *Physical Control of the Mind. Toward a Psychocivilized Society*.[22] The large majority of Delgado's research in the late 1950s and early 1960s may be summarized as a period of cataloging as many demonstrations of the effectiveness of ESB for the direct modification of behavior as possible, and his research was funded by the Office of Naval Research and the US Air Force. It is interesting to note that the neuroscientist and historian, Prof. Elliot Valenstein, had characterized this period of time as one during which efforts were made "to describe as many behaviors as possible that could be elicited by brain stimulation. As the list of responses that had been elicited grew larger, the potential for controlling behavior and motivation seemed greater and greater. Often there was uncritical acceptance of the significance of these reports, which characteristically claimed to have tapped an anatomical circuit regulating specific emotional [or drive] states."[8]

Although Valenstein argued, in his 1973 review of this work,[8] that there was a vast leap to be made from these non-human primate demonstrations of ESB to the development of human clinical applications, there were several clinicians at that time who were making such attempts with great enthusiasm. Chief among these researchers was Robert G. Heath and his team at Tulane University. Heath's ESB program was an off-shoot of the Columbia-Greystone Project, started in the 1940s, with the goal of discovering the central nervous system etiology of schizophrenia.[23] Between 1950 and 1960, Heath's team had implanted large numbers of electrodes in the brains of 19 schizophrenic patients, many using the "open method" for placement of electrodes directly into the septal region, the amygdaloid bodies, the thalamus, and other regions of interest. By 1952, stereotactic operative techniques had been adopted at Tulane, and "any one patient may have had several dozen or more electrodes" implanted throughout the brain.[23] From this first series of patients, four individuals suffered iatrogenically induced seizures, six patients suffered infections ultimately resulting in two deaths, and one patient suffered acute cardiac failure.

By 1960, Heath's team had conducted ESB experiments on 52 patients (42 schizophrenics, 6 with intractable pain, and 4 with medically refractory seizure disorders). Some of the ethics issues related to this program of research, including questions related to the lack of appropriate informed consent (even by widely accepted standards in the 1950s), have been well described elsewhere.[23] As for the later history of this program, with the discovery of chlorpromazine in 1955, these types of programs that focused on somatic treatments for psychosis (e.g., ESB) rapidly declined. By the early 1960s the emphasis in Psychiatry had shifted to the growing field of psychopharmacology, and as a result, between 1960 and 1970 only eight new patients were treated with ESB at Tulane.

Despite the extraordinary efforts made by this large program in Louisiana, which had applied ESB to treat psychiatric disease, chronic pain, and epilepsy in human patients throughout the 1950s and 1960s, Delgado offered scant mention of this work in his own writings. In fact, Delgado's 51-page published transcript of his invited *James Arthur Lecture on the Evolution of the Human Brain*, at the American Museum of Natural History in New York City in 1965,[21] makes only a single reference to Heath's work. I was also not able to find a single reference made to the similar studies in human patients conducted at around the same time by Drs. Vernon Mark (Boston City Hospital) and Frank Ervin (UCLA), including collaborative work with Delgado.[15] Moreover, in Delgado's 1969 book on this subject, he cites 16 of his own published articles (most of these being single-authored), only three papers by R.G. Heath, and no articles at all by either Mark or Ervin.[21] It appears that the ESB field in the late 1960s, and especially the translation from animal models to human clinical applications, was an intensely competitive area to be working in, and Delgado strived to protect the "primacy" of his own contributions.

By 1963, Delgado was an Associate Professor at Yale and in the middle of an illustrious career, having just won a prestigious Guggenheim Fellowship. However, this period of his career coincided with a waning interest in the use of invasive surgical techniques to treat psychiatric illness and it was critically important for him to maintain high visibility for his work, in order to retain funding from the Department of Defense, to retain a large amount of laboratory space and staff at Yale, and to ensure that his discoveries would be assured a continuing place on the cutting edge of biomedical advancement. As noted above, Delgado had a reputation amongst his colleagues and students as a man with a "healthy ego," and he routinely scripted, directed, and narrated mini-documentary films of his own making, to record the achievements made in his laboratory. Thus, it is not surprising that, in response to what was likely to have been his perceived desire to show the public that his work was exciting, promising, and imaginative, that he would choose to choreograph and film one of the most dramatic scientific demonstrations in the modern history of the neurosciences.

2 Delgado's brave bulls "experiment"

In 1963, Delgado completed a demonstration experiment to prove to a skeptical public, and to the scientific community, that he could turn a brave, angry, and dangerous beast into a docile animal with the push of a button. On a bright spring afternoon, Delgado, both academician and performer, stepped into a Spanish bullring unarmed, so that a "brave bull"[c] would charge at him with all of its "destructive fury." As this exercise was not an actual experiment in the traditional sense of the word, Delgado's demonstration of his ability to stop the bull before being impaled, by use of a hand-held transmitter that sent signals to electrodes buried deep within the animal's brain, was intended for a single purpose – to make headline news back home in the United States, and his effort proved to be quite successful.

On May 17, 1965, two photographs and a story appeared on the front page of *The New York Times*, setting the scene for a disbelieving readership: "The afternoon sunlight poured over the high wooden barriers into the ring, as the brave bull bore down on the unarmed matador – a scientist who had never faced a fighting bull." *The New York Times* science writer, J.A. Osmundsen, concluded his report by stating that this was "…probably the most spectacular demonstration ever performed of the deliberate modification of animal behavior through external control of the brain."[24]

The purpose of this dramatic demonstration was unmistakable, as Delgado personally scripted and narrated the 8 minute, 45 second documentary film produced to record this event. Delgado introduces the film with the reminder that "The brave bull is a proud and beautiful beast … that may attack with destructive fury." He continued his narration of the film with the clear message that "The purpose of our investigation was to see whether the genetically determined aggressive instinct could be modified by ESB." This film, and more commonly still photographs obtained from the raw footage, have been reproduced in numerous history of neuroscience textbooks, and even contemporary magazine articles for the lay public.[13] In fact, a short segment of the film, with a narrator other than Delgado providing the "voice over," is currently posted on the popular video-sharing web site, YouTube.com. To my knowledge, however, the entire original film from Delgado's laboratory has not been shown, in its entirety, in many, many years.

My father, Dr. Daniel R. Snyder, served as Delgado's last American-trained post-doctoral fellow at Yale in the early 1970s, before Delgado returned to

[c]Perhaps fortuitously for Dr. Delgado, the anthropomorphic use of the term "brave" to describe the bulls used in his demonstration was actually factually quite accurate – despite the typical use of the word to describe someone who is fearless in the face of danger. The specific breed of bull used for bullfighting, and for his demonstration, was the "Brave Bull."

Spain in 1974 and Snyder took over Delgado's monkey colonies and laboratory space at Yale for approximately 10 years. As a result of this family connection to Delgado, a number of items (e.g., books and several reels of 16 mm film) made their way from the laboratory to my parents' home). The original print of the "Brave Bulls" film was found on the top shelf of my childhood bedroom closet, where it most likely resided under a pile of T-shirts for 20 years or more. With generous financial assistance from Pfizer Inc., I worked with a local television editor in New London, Connecticut, to clean, re-master (to correct slippage between the video and voice-over narration), and digitize the original print of this film. ***This nearly 9 minute film, written and narrated by Professor Delgado himself, is included as a CD insert with this volume, and I encourage all to view this film themselves and to evaluate the veracity of my conclusions provided below.***

In reviewing this film myself, I have made the assumption that, since Delgado clearly exercised creative control over the content of the finished product, that he most likely selected his "best evidence" to support his stated central aim for the demonstration, that is, to modify aggressive behavior. The film describes the placement of 21 electrode points, along several leads, placed stereotactically into the brains of two bulls at a breeding ranch in Cordoba, Spain. In the film he mentions three specific regions of interest in the brain that were individually stimulated. First, he noted that stimulation of the rostral thalamus did *not* modify aggressive behavior. Next, stimulation of the internal capsule (within the striatum) was reported to result in stereotypic turning by the animals. However, Delgado spends most of the time in this film discussing the results obtained with electrical stimulation of the caudate nucleus. Specifically, Delgado noted that before stimulating the caudate nucleus "aggressive behavior against the cape is evident ... without radio control the animal is rather unfriendly." By comparison, following ESB in the caudate nucleus, "the animal is induced to stop and turn away ... then spontaneous aggressiveness is tested once more."

This last remark, made on the film, is a puzzling one. Did Delgado mean to imply that stimulation of the caudate did anything more than cause a stereotypic contralateral turn by the animal? Did it also lead to a temporary cessation of aggression? After watching this film many times, and after reading Delgado's published descriptions of this event, I am left with the belief that Delgado may have been intentionally vague on this crucial point. There are several reasons why I suspect that this message was intentionally left difficult to interpret clearly by Delgado. First, the most important goal for this demonstration was summarized on a single sheet of typed paper presented at the very end of the film. As it is difficult to read in the film itself, the verbatim transcript is as follows: "*The purpose of this research is to examine the striking difference in the conduct of brave and peaceful bulls and to investigate the structures of the brain which may be responsible for the hereditary transmission of aggressive behaviour.*" If Delgado had merely shown that stimulation of the caudate nucleus led to either motor arrest and/or involuntary turning by the animal, then this demonstration would fall far short of its mark – and it would have in all likelihood not have made the front page of *The New York Times*.

Second, the ambiguity of this message continued with his later publications. On page 32 of his published 1965 invited address at the American Museum of Natural History in New York City,[21] Delgado related the results of the Brave Bulls demonstration to laboratory work with rhesus macaques, in noting that stimulation of the caudate nucleus leads to the inhibition of aggression – however, the actual behaviors described are solely motoric ones. Moreover, in his 1969 book,[22] intended for the educated lay public, Delgado seemed to confuse this point again by stating that "When the caudate nucleus is stimulated, the normally ferocious macacus rhesus becomes tranquil, and instead of grabbing, scratching, and biting any approaching object, he sits peacefully and the investigator can safely touch his mouth and pet him." But, again, is this the control of aggression in the animal with caudate stimulation, or more generally an arrest of intentional motor behavior?

It is clear that at least a few of Delgado's contemporaries were equally puzzled by what appeared to be mixed messages from him on this point. As an example, a published verbal exchange that followed a 1965 symposium presentation by Prof. Irenäus Eibl-Eibesfeldt (Max-Planck-Institut für Verhaltenphysiologie, West Germany), between Delgado and several participants at the meeting, highlight the disagreement regarding the impact of caudate nucleus stimulation of modifying aggression *per se*. Delgado was an attendee of the 5th Conference on Brain Function, held in November of 1965, and sponsored jointly by the Brain Research Institute at UCLA and the US Air Force Office of Scientific Research. Other notable attendees at this conference included Eibl-Eibesfeldt, Donald B. Lindsey, Carmine D. Clemente, Pierre Gloor, Louis S.B. Leakey, and Karl H. Pribram. In an exchange that followed Eibl-Eibesfeldt's talk within the published conference proceedings,[25] Delgado mentioned his recent demonstration in Cordoba, Spain, with a nearly verbatim transcript of the documentary film reprinted in the published proceedings from this meeting.

Following Delgado's narrative, Dr. Bryan Robinson (Yerkes Regional Primate Center, Atlanta, Georgia) asked, "Dr. Delgado, did you try stimulating the animal while it was either eating or mounting, to see if the effect were specific to aggressive behaviour?" This question led to an exchange in which other participants raised similar concerns. Dr. Irwin S. Bernstein (also from the Yerkes Regional Primate Center in Atlanta) asked a very similar follow-up question of Delgado, "Were you stimulating an inhibition, or another response which interferes with aggression?" Delgado responded by stating "I think we can do both, depending on the cerebral structure stimulated." Dr. James Hedlund (US Department of the Army) made a third attempt to gain some clarity on this point by asking, "Although you have not specifically invoked neural mechanisms as the inhibitory mechanism, by the notion of continued inhibition over, say, 5 minutes, I presume you refer to neural inhibition. But I would like to ask if the brain stimulation and abrupt interruption of ongoing behavior had enough noxious characteristics to condition the animal against any response that was taking place, and in this case aggressive behavior. Was it learned as opposed to neural inhibition?" This question led to a continued exchange between Delgado and Hedlund, in which there was clear disagreement regarding whether the direct stimulation of the caudate nucleus in the bull could lead to a cessation of ongoing activity that might, in itself, be noxious and would lead to a negative reinforcement with regard to aggressive behavior.

Interestingly, at the end of this exchange, Delgado promises that, "I do not think this is the case; I shall present more experimental data to support my point later during this conference." In his own presentation at the conference, Delgado discussed his laboratory research involving ESB of several regions of interest in the cat and monkey models, including stimulation of the caudate nucleus. With regard to stimulation of the caudate, he reported his findings that submissive male macaques in a social colony will learn to press a lever to induce an ESB to the caudate nucleus of the dominant male in the colony, with the result being that "caudate nucleus stimulation [produced] inhibitory effects, including a decrease in offensive–defensive reactions."[26] To be fair to Delgado, he also presented data to suggest that, when the caudate nucleus of the dominant male in the colony was stimulated, there was no resulting modification of "nestling, balling, lying down or being groomed, but [caudate stimulation] inhibited aggression so effectively during the period of excitation that the boss could safely be touched with bare hands and easily caught inside the cage." In truth, with the rich afferent and efferent connections between the striatum and limbic structures, it is possible that direct and focal caudate stimulation could lead to an arrest of behavioral aggression. The problem, however, is that with the data that Delgado had in hand, it was just not possible to disassociate the effect of caudate stimulation on aggression from the effects of the same stimulation on basic motor control. I suspect that Delgado knew that anything less than a very clear and unambiguous story would not make the front page of *The New York Times*.

3 Electrical stimulation of the caudate nucleus

One might reasonably question whether Delgado was fully aware of the published physiological research that by 1964 had repeatedly demonstrated the effect of the caudate nucleus stimulation on the inhibition of movement. By the early 1960s there was already a sizable literature demonstrating this point. For example, in 1957, Forman and Ward reported on their studies with domestic cats, in which they stimulated 66 separate points within the head of the caudate nucleus. The most commonly elicited response with stimulation was a forced contralateral head turn, with the inclusion of contralateral foreleg and hind leg extensions with stimulation of 20–44 separate points located in the caudal half of the head of the nucleus.[27] Similarly, in 1964 McLennan and colleagues stimulated the caudate nuclei of domestic cats and found that the effects of electrical stimulation depended on the intensity and frequency of the electrical impulses, such that with low frequency and intensity pulses there was a complete arrest of motor behavior observed, and with greater frequencies and intensities of stimulation a forced contralateral turning by the animal was induced.[28] The images in this published paper of the cats being induced to turn in a tight circle, in the direction contralateral to the side of stimulation (the higher the frequency of stimulation, the faster the turning behavior), look strikingly similar to the film images of Delgado's bulls during ESB of the caudate nuclei. What McLennan and colleagues described in the cat,[28] and what we see on the film of the bulls by Delgado, is a forced, stereotyped motor response to stimulation of motor centers in the *corpus striatum*. If there were also an effect of this same stimulation on the mediation of aggression, then how could this be meaningfully discerned, even by the trained observer? And, if the necessary double dissociation of these separable but non-exclusive explanations had not been teased apart by careful experimentation at the time, then surely the complexity of the story would not make for easy reading by the public in national news outlets.

Another question to be raised is, if Delgado knew that stimulation of the caudate nucleus would lead to an arrest of motor behavior and/or stereotypic contralateral turning, and that these behaviors would be difficult to clearly dissociate from possible effects of the same stimulation on the expression of "aggression," then why did he focus his demonstration on the caudate nucleus in the first place? Why not choose other regions of interest (e.g., several areas within the limbic system) that would be more likely to directly affect the emotional valence of the animal, without concurrent and dramatic alterations of motor behavior? Although we cannot know the answer to this question for certain, my suspicion is that the answer is quite obvious. Imagine, if you will, that you were about to face a 500–800 kilo charging bull with nothing more than a radio transmitter in your hand. Would you prefer to attempt to alter the emotional state of the raging bull that is charging toward you at full speed, or would you prefer to force the bull to "stop on a dime" and rapidly turn away from you? And yet, which of these two alternative choices would lead to the more sensational story?

4 The reporting of the Brave Bulls demonstration

Despite the obvious significance of this demonstration of "brain control," there was a substantial misunderstanding in the reporting of this story. The wide readership of *The New York Times* were led to believe that the significance of Delgado's work was in his purported ability to control, manipulate, and eradicate violent behavior in an otherwise aggressive animal. By implication, this work held tremendous significance for the control of human aggression and related criminal behavior. However, Delgado was most likely aware that he was not directly stimulating a brain region considered to play a central role in the modulation of affect. Rather, he had placed his electrodes into the caudate nucleus of the bull, a part of the brain that at the time was already well known to be involved principally in the mediation of motor learning and coordinated motor behaviors, but was not considered to be central part of the brain circuitry underlying the modulation of "emotion" or "aggression." *Delgado did believe that the stimulation of this particular brain region led to reduced aggression in animals*, as well as affecting its more primary role in modulating coordinated movement. He was also likely aware that the ability to remotely control intentional movement by the bull would be seen as less impressive than his ability to modify emotional tone, the latter having far more important ramifications for potential treatments of human psychiatric diseases. His emphasis on the significance of caudate stimulation for the modulation of aggression, particularly when conveying the significance of his research to the reporters covering this story, is quite understandable but still striking, and this famous episode in the history of 20th century neuroscience epitomizes one of the central themes of this volume. What are the ethical responsibilities of scientists with regard to their reporting of novel research? How should scientists in, say, Delgado's position, balance the desire to tell an exciting story with the need to exercise caution when the results of their work are not yet fully understood?

Delgado knew well that his facing a charging bull, with nothing more than a small radio transmitter to send signals to electrodes implanted in the animal's brain, would capture the imagination of a public hungry for scientists to lead the way to curing terrible human diseases. In reflecting on this event a few years after it occurred, the neuroscientist and historian, Elliot Valenstein, wrote:

> We are living in a period of such rapid change that the average person today believes almost anything is possible and, as a result, science fiction descriptions of the future are often accepted as realistic predictions based on firsthand knowledge of the subject matter. Actually, the accounts of brain stimulation read by most people are typically written by free-lance writers or imaginative social scientists, who have no direct experience with the techniques involved and are often uninformed about even basic brain physiology. Nevertheless, the possibility of controlling the moods and behaviour, if not the thoughts of individuals by remotely controlled electronic devices has such obvious dramatic possibilities that the topic is virtually irresistible to those with a more speculative bent.[8]

These words are as relevant today as they were nearly four decades ago. A recent article in the popular US magazine, *Scientific American*, described Delgado's life and work – including the Brave Bulls demonstration (with still images from the film), and the results of that research were described as *both* halting an intended motor behavior (the bull's charging toward Delgado in the bullring) and modifying the aggressiveness of the animal at the very same time.[13] It is worth noting again that his Brave Bulls demonstration was not a classic controlled experiment, and his interpretation of the observed results were – at the time of the event – clearly open to debate by his peers. In fact, although there are two publications in Spanish (one of them in a lay publication, *Gaceta Ilustrada*),[29,30] I have not been able to locate a single peer-reviewed English language publication detailing this demonstration in the scientific literature, but rather, he often mentioned this event in non-peer-reviewed book chapters and invited addresses.[21,22,26]

5 The urge to simplify complex and confusing research

What were Delgado's motives for simplifying his research results, and the complexities of the phenomena he was exploring, for public consumption? Delgado was an exquisitely talented surgeon, a skilled neurophysiologist, an innovative inventor, and he probably deserved at least one honorary graduate degree in electrical engineering. By the time he returned to Spain in 1974 to aid in establishing a new medical school in Madrid, he had received the *Gold Medal Award* from the Society of Biological Psychiatry, and the *Rodriquez Pascual Prize* from the Government of Spain (awarded to a scientist in the Spanish-speaking world whose medical research has global significance). Perhaps Delgado was smitten by the potential power of his techniques, and their applications, along with the science writers who chronicled his work. In a 1970 article about him, in the *Yale Alumni Magazine*, Karen Waggoner wrote that Delgado's "...personal research goals, he admits, are not at all modest. 'I would like to cure epilepsy, cure mental disturbances, and construct a better world. That's all,' he says with only a trace of a smile."[31]

Although I do not believe that Dr. Delgado consciously deceived journalists about the proper interpretation of the Brave Bulls demonstration, he did little – at the time of the reporting of this event – to ensure that the real complexity of the story was apparent. Five years later he wrote that the result "seemed to be a combination of motor effect, forcing the bull to stop and turn to one side, plus behavioral inhibition of the aggressive drive. Upon repeated stimulation, these animals were rendered less dangerous than usual, and for a period of several minutes would tolerate the presence of investigators in the ring without launching any attack."[21] This statement may or may not actually be a valid one, but assuming that it is, his initial efforts to communicate the science behind his work to journalists (and in the film) clearly contained errors of omission. Why? Clearly Delgado enjoyed the public spotlight, and what better way to encourage such attention than by staging a dramatic demonstration with a clear and simple message that captures the imagination of a public eager for the transmutation of "science fiction" into science fact? As noted above, my father was the last American post-doctoral fellow to be trained in Delgado's laboratory before he returned to Spain in 1974. In reflecting back on his mentor, he noted that "...there were at least four Delgados: (1) the one writing or co-authoring for peer-reviewed scientific journals; (2) the one being interviewed for the lay press; (3) the one re-writing his stuff (over and over) for re-publication overseas in several languages; and (4) the one writing as a social philosopher – as in his 1969 book. I think that it is pretty clear that the film narration falls under Delgado #2, the one playing to the tune of an enthusiastic public."[32]

6 The social impact of misrepresenting scientific results

One might argue that Delgado's over-simplification of his interpretation of the results of the Brave Bulls demonstration, for public consumption, is really a minor oversight with little consequence. There are two reasons why I think that this is not necessarily the case. First, Delgado unquestionably stood to gain from the presentation of the clear message that he had discovered the means to modify aggression. He wanted to convince his funding agencies that he had developed the means to exert such a powerful effect on behavior, and scientists within those agencies were not sure themselves that this was what was indeed being demonstrated.[25] Further, the increased notoriety following this event would be a boost to his career, his place within the academic structure at Yale, and his position as the preeminent leader in his field in the eyes of his contemporaries. Hence, there were several potential sources of secondary gain for Delgado – all enticements to omit some of the detail, the richness of the story, when interacting with journalists and in the preparation of his famous film (provided with this book for you, the reader, to view and consider).

Second, and more broadly, this vignette taken from the mid-20th century speaks to a larger theme that is revisited several times in this volume – that is, the uneasy partnership that exists between biomedical scientists and the media. Thompson and Nelson note that, "In their search for newsworthy stories, journalists seek what is new, interesting, and unexpected."[33] As Laura Spinney notes in her essay (Chapter 13), these motives on the part of science journalists stand in contrast to what the scientist often hopes to present; a body of work that is cumulative and complex. It is often the case that whereas journalists report information that is practical and immediately relevant, scientists are trained to report new knowledge to their peers that may be incomplete and full of unanswered questions. In the case of many scientists across the history of modern biomedical science, including Louis Pasteur in the late 1880s (see Chapter 2, by Prof. J.H. Warner) and Delgado in the 1960s, as merely two examples, they often benefit from media attention to their work because it "provides public and professional recognition, circumvents the lengthy lag times of scientific journals, and draws attention of policymakers to their research area, which potentially enhances research applications and future grant funding."[33] These are, I think, some of the motives that led to the manner in which Delgado reported his work, why he chose to step onto the world stage with such a spectacular piece of scientific theater, and why he engendered some of the criticism and disparagement that he has over the years.[8]

CHAPTER

4

Adrenal Transplantation for Parkinson's Disease

Dennis D. Spencer

Our nervous system shares susceptibility with all other organs to some common pathologies, such as cancer and infectious disease. However, it is often uniquely disrupted by focal cell loss that may represent less than 1 percent of the total cells comprising our brain, spinal cord and peripheral nerves. This impact comes from the layered heterogeneous, functional specificity of the cell groups and the fact that they work in a bioelectric network that when activated during our development produces the sentient being that defines you. Thus, very focal spinal cord trauma, head injury or stroke may leave one paralyzed and cognitively impaired. Degenerative diseases such as Alzheimer's, movement disorders, multiple sclerosis (MS) and amyotrophic lateral sclerosis (ALS), destroy our function over a variable time but nervous system injury and degeneration have one thing in common – they ultimately destroy who we are.

It should not be surprising then, given the focality of impairment, that physicians and scientists should look to replacing these lost cells as a restorative answer to the problem. Likewise, patients who see inevitable functional loss are eager to embrace even limited promises of cure.

Biomedical research has not provided selective prevention or cure for many diseased and degenerative organs including the brain and spinal cord. However, as is often the case when science lags behind technological advances, major advances in surgical techniques and perioperative care stimulated a steep rise in solid organ transplantation especially during the 1970s and 1980s. A governmental Office of Technology Assessment (OTA) report in October 1990 noted that in 1989, over 500 000 Americans received organs or tissue transplantation including bone, cornea, kidney, liver, heart, pancreas, lung and bone marrow.[1] This same report which concentrated on the state of neural grafting also emphasized the tens of millions of Americans with neurological diseases and the greater than $100 billion per year in medical expenses and lost income from the neurological diseases most amenable to replacement of lost cells affecting all age ranges, led by Alzheimer's, followed by stroke, CNS injury, epilepsy, movement disorders, MS and ALS.

Tissue grafting in the CNS would make sense for several strategies, some to replace neurotransmitters or chemicals, to stimulate survival and growth or to actually replace structural elements in the network.

Research in this arena actually began in the 19th century and the first attempted CNS tissue transplantation in animals is attributed to American scientist W.G. Thompson in 1890, and the first successful grafting was of neonatal CNS into an adult rodent brain, performed by E. Dunn in 1917.[1] However, experimental advances in the field were slow until 1980s. Interest then grew rapidly and peer reviewed publications, which had numbered less than 10 in 1980 doubled about every 2 years when these figures were reviewed in 1988.[1]

There were many factors that accounted for this acceleration but chief among them was a perceived rise in the prevalence of Parkinson's disease, the realization that medical therapy was time limited, the clear understanding of what specific cells were lost and the appearance of reliable animal models for the disorder.

James Parkinson described a unique group of patients with shaking and impaired mobility in a 1817 monograph called "the shaking palsy."[2] These symptoms were later found to be secondary to the selective progressive death of dopamine-producing cells in the substantia nigra of the brain. Dopamine cannot be replaced directly because it does not pass through the blood brain barrier. However, in 1959, it was discovered that a precursor of dopamine, L-Dopa, could pass through this barrier, be metabolized by remaining substantia nigra to dopamine and thousands of patients had symptomatic relief. However, the neurochemical replacement does not stop the progression of the cell loss and symptoms eventually return with the patient's condition worsening. Approximately 1 million individuals suffer from Parkinson's disease in the United States with 50 000 added each year. The average age of onset is around 60 and duration until death ranges from 8 to 14 years.

Although the majority of patients with the constellation of symptoms constituting Parkinsonism have the idiopathic disease, the cause of which is unknown, certain environmental toxins are known to result in the same cell loss and array of symptoms. In 1982, a number of young heroin addicts had injected a home-made narcotic, which contained the impurity, 1-methyl 4-phenyl 1,2,3,6-tetrahydropyridine (MPTP).[1] They all quickly developed the symptoms of Parkinsonism, and it was discovered that MPTP caused selective loss of substantia nigra cells. Investigators immediately took advantage of this and created a variety of animal models that were used in transplantation research.

Another relatively robust animal model of Parkinson's disease was the 6-hydroxy dopamine (6-OHDA) lesioned rodents.[3] Using both of these models, investigators searched for appropriate tissue that would either replace lost substantia nigra cells or at least provides Dopamine as a cellular product.[4] Fetal midbrain tissue was used successfully in both models, the tissue surviving, making Dopamine and reversing deficits.[5] Ethical controversy and immunological obstacles limited the early utilization of this method and the search for alternative tissue revealed the adrenal gland as a reasonable substitute. Chromaffin cells separated from the gland survived grafting and produced dopamine in amounts that also reversed movement abnormalities in rodents.[6] As autologous tissue (tissue from ones own body), the adrenal gland avoided the ethical and immunological problems and, therefore, in 1982, Backlund stereotactically implanted cell suspensions into the Putamen of two patients with Parkinson's disease.[7] In a 1985 report, no improvement was noted other than a transient improvement in "on" time in one patient.[7] Despite these negative results, Ignacio Madrazo and colleagues chose two young (35 and 39 years of age) patients with Parkinsonism and implanted their own adrenal chromaffin tissue into the right caudate nucleus via an open craniotomy and transcortical approach.[8] A 3 μm^3 cavity was dissected in the caudate and the medullary tissue implanted and anchored with stainless steel staples such that tissue fragments were in contact with both the nucleus and the Cerebral Spinal Fluid (CSF). After 15 days, they stated that "functional recovery occurred on an almost daily basis" in patient number one and improvement continued for the 10 months reported. Patient 2 improved almost immediately and 3 months from surgery he, as patient number one, had no tremor, no rigidity, clear speech and normal facial expression. This 10 and 3 month respective follow up was published in *The New England Journal of Medicine*, one of our most prestigious publications. In the discussion, Madrazo and colleagues speculated that the ventricular placement appeared ideal, that the bilateral improvement noted must be due to circulating dopamine from the adrenal tissue and not caudate injury and that this procedure results in "amelioration of most signs of Parkinson's disease."[8]

And so, this dramatic amelioration of Parkinson's disease moved from the pages of our most respected medical journal to one of our most respected newspapers, *The New York Times*, in about the time it took the digital ink to dry.[9] The headlines read "Transplant Brings Dramatic Gains for 2 Parkinson's Victims." The reporter, Walter Sullivan, either not realizing the follow-up was very short, or disregarding the implications of this and the unexplained almost immediate bilateral improvement, wrote with the usual journalistic style emphasizing the dramatic outcome and even found a scientist at SUNY (Stony Brook) who called the report "an important event in the history of the treatment" of the disease. Indeed, Dr. Moore, The SUNY scientist, wrote an editorial in *The New England Journal of Medicine* which accompanied the Mexican article, urging the National Institutes of Health (NIH) to "organize coordinated clinical trials." The perplexing nature of the Mexican results was not mentioned until the last sentence of the article.

A few days later, Vanderbilt neurosurgeons notified the *Times* of a similar operation that was to take place on April 10, 1987, and the newspaper was there reporting the event that same day as the first adrenal transplant for Parkinson's disease to be performed in the United States.[10] In this case, the human experiment itself, not a peer-reviewed publication of outcome, was disseminated worldwide and the impact was remarkably predictable, not only on the hundreds of thousands of patients with the disease but also on the physicians eager to be early adopters of such a dramatic therapy. Many likened the reports of transplantation surgery for these patients to the discovery of Penicillin. The evolving excitement was picked up by Time Magazine on July 13, 1987 following the Vanderbilt transplants where they concentrated on the personal story of a 39-year-old patient's apparent remarkable recovery only a few weeks after her surgery. Although some physicians and scientists interviewed at a conference by the *Times* reporter, Leon Jaroff, expressed concerns about lack of long-term outcome and the paucity of data explaining the quick recovery, the conversation soon turned to expansion of the process to Alzheimers, Huntingtons and other neurodegenerative diseases. The article ended not with healthy skepticism but with multiple centers ready to move forward with individual small studies and no one willing to be left out of the competitive rush to a potential cure of this dreaded disorder. The patients would do anything for a chance of a better life, and the very positive twist that the media applied to these anecdotal stories resulted in a clamor for more, the media complied and the medical community was soon caught up in the rushing river initiated and endorsed by a handful of its membership outside the typical scholarship and constraints of human investigation. Scientists asking for solid clinical researchers to paddle against the current and wait for more data or for a controlled trial were ignored. Instead, many of their colleagues, caught up in the enthusiasm and excitement of the moment, paddled forward as fast as possible and the heady, competitive race down stream to the precipice was on.

There was to be no randomized clinical trial, perhaps because even if NIH had the mechanism to rapidly originate and support such a trial, the logistics of a multicenter collaboration defied the entrepreneurial spirit of individual neurosurgical programs and their collaborators. Time was wasting and desperate patients were lining up. Less than a year after their first report, Medrazo *et al.* reported similar positive findings in 11 patients now ranging in ages from 35 to 65.[11] In two short years, approximately 300 operations had been carried out worldwide and in most, neurosurgeons attempted to emulate the Mexican procedure. By this time, the sporadic early cases operated on in the United States were being reported, as registries were formed to provide collective *post hoc* outcome data. The first 6 months' data of one of these registries noted only modest improvement in 19 patients with a reduction of off periods (when a patient suffers the most severe symptoms). The patients remained disabled and on all of their medications. Significant morbidity was also noted, consisting of respiratory problems, urinary tract infections and disorders of cognitive function, (confusion) consciousness, and psychiatric problems including delusions and hallucinations.[12] This period of transplant fervor was also marked by multiple publications emphasizing not outcome but variations in the technical aspects of the yet unproven procedure, including best methods (retroperitoneal) for harvesting the adrenal gland, how to dissect it, and how to deliver it to the basal ganglia using such techniques as dividing a small section of the corpus callosum (the fiber bundle that connects the two cerebral hemispheres), utilizing a fibro-optic endoscope through the frontal lobe or reversion to the stereotactic techniques employed by the original Swedish investigators which was safer but ineffective.[13–17]

By 1991, as noted above, the US registries were reporting unsustained long-term improvement.[18] However the Mexican group, in a 1991 publication called "Autologous Adrenal Medullary, Fetal Mesencephalic and Fetal Adrenal Brain Transplantation in Parkinson's disease: A Long-Term Postoperative Follow-up," argued that four of their original five patients had maintained their initial improvements for 3 years. They then concluded "regardless of the mechanisms of action involved, these brain transplantations are effective procedures to ameliorate PD signs to an extent not achieved by any other means."[11] Peer reviewers appeared to be convinced by their data and to agree that the only challenge now was to select the best candidates for surgery and the best donor tissue.

Some lay articles during this period were better balanced, although even those that attempted to demonstrate the controversy that Madrazo had unleashed managed to provide a forum for his unwavering optimism about the the procedure and his leadership position in this new field. Larry Richter in a 1988 article for *The New York Times* leads off with a paragraph describing Madrazo in his office surrounded by letters from patients pleading for him to operate on them. He then goes on to describe how some investigators are questioning his results with the following quote: "We're fed up with his being lionized," said Judy Rosner, executive director of the United Parkinson Foundation in Chicago, reflecting the hard feelings about Dr. Madrazo's claims that have arisen in some quarters. "It's so darn dangerous and expensive what he's doing. And besides, we think some of his patients don't even have Parkinson's, only disorders that look like Parkinson's."

When the controversy continued to roil with his purported additional positive outcomes with fetal cell transplants, Madrazo replied with the following quote: "It's very much like what we saw after Christian Barnard did the first heart transplant," Dr. Madrazo said. "Those who got good results were delighted, and those who did not grumbled. But after a while, everything finds its level."[19]

Finally the fabric of success and recognition began to unravel from the US registries and also from publications of other Mexican Center, but not before over 300 patients had undergone implantation in uncontrolled heterogeneous trials with unsustained positive effects and many complications.[20–23] In a Progress in Brain Research chapter entitled: "Is autologous transplant of Adrenal medulla into the striatum an effective therapy for Parkinson's Disease," E. Garcia Flores *et al.* reported on 24 patients implanted identically to the Madrazo study. They found no evidence to support claims of improvement in appropriately diagnosed Parkinson's patients. There was an immediate improvement in bradykinesia and tremor starting as early as 12 hours, lasting for 2–3 weeks and a second response, which waned at 6–8 months. They felt that the transient improvement could be explained based on one or all of three explanations, which had been published in the literature. The first related to injury to the caudate nucleus, the second to short-term release of dopamine by the medullary tissue and finally to an inflammatory response of the injury. Regardless, they concluded: "that autologous medullary grafts have modest, if any, beneficial effects in Parkinson patients, and that the operation has evident risks that outweigh the questionable benefits. For this reason, we performed the last such operation on the 17th August 1988."[24]

Individual small group studies such as the one just described continued to show a modest improvement, if any in their patients but the Madrazo report had already stimulated the approximately 300 operations worldwide. Finally, because of the lack of a controlled trial, two registries were formed in the United States. The first was collaboration between the American Association of Neurological Surgeons (AANS) and the National Institute of Neurological Disorders and Stroke (NINDS). They established the General Registry of Adrenal–Fetal Transplantation (GRAFT) and together with another registry set up by the United Parkinson Foundation they acquired ongoing data from groups in the United States and representative teams from Canada, Mexico, Europe and South America. Approximately 200 patients were presented in the Neural Grafting report, confirming the limited efficacy of the adrenal grafts and clearly establishing major discrepancies between the original Madrazo trial and ensuing studies.[25]

The adrenal transplantation for Parkinson's disease is an all too common aberration in the usual cycle of disease-based research. Most commonly, research takes place as a low-frequency background waveform superimposed on a graph where the Y-axis is "discoveries leading to a cure of the disease sometime in the future" and X is "time." Although this relationship is not necessarily linear over, that is a gradual accumulation of new discoveries rising until the problem is solved, in fact, we are somewhere along that curve in all diseases afflicting humans. Sudden discoveries that change the paradigm of a disease such as antibiotics and chemotherapy for certain forms of cancer give us hope that around the corner the cure of our disease interest lies waiting for us to be the miracle worker. It is the hunger that we as physicians and scientists share with our patients that make us both vulnerable to self-deception. That self-deception can be magnified by the individuals who provide the link between the physician, scientist and patient, that is, the editors, journalists and peers who may buy in quickly to a seductive solution. In a typical appropriate clinical investigation, the hypotheses are considered, as Popper thinks they should, with an eye to disprove the theory and most often the rule is that only pieces of a discovery are validated and move us upward. When this rule is broken and the hypothesis is accepted at face value, you are almost always wrong and you have violated the rule of healthy skepticism. In the case of Parkinson's disease we are more susceptible to this violation because we are farther along the curve than with many disease processes but also at a dangerous place in that curve. We understand the mechanism of the disease but we don't understand the pathophysiology, for example, we understand that a certain cell loss results in certain changes in biology that explain the disease's behavior. Unfortunately, we do not understand why those cells are lost, what initiates the process, what sustains it and what the underlying molecular genetic environment interaction is. Because we can dramatically affect the disease's behavior temporarily by replacing that final downstream molecule, Dopamine, in a variety of forms, first as a drug (L-Dopa), then as cellular replacement, the world anticipates a cure is at hand. In our example here almost everyone fell into the trap from the initial investigator, Dr. Madrazo, who I'm sure was well intentioned and saw a clear black and white solution using the tools of the time. Perhaps he was not surrounded by enough local skepticism, but surely he was not solely to blame when his premature findings were embraced by his peers and launched to center stage by one of our most prominent medical journals, *The New England Journal of Medicine*. That moment of recognition was bound to be illuminated more by *The New York Times* and the course was inevitable with the cries of desperate patients and surgeons eager to answer them.

Peter Fonagy, Ph.D., D.Psych., F.B.A.
Freud Memorial Professor of Psychoanalysis
Director, Sub-department of Clinical Health Psychology
University College London, London, UK
Chief Executive Office, Anna Freud Centre, London, UK

Peter Fonagy, Ph.D., D.Psych., F.B.A., is Freud Memorial Professor of Psychoanalysis and Director of the Sub-department of Clinical Health Psychology at University College London; Chief Executive of the Anna Freud Centre, London; and Consultant to the Child and Family Program at the Menninger Department of Psychiatry and Behavioral Sciences at the Baylor College of Medicine. He is Chair of the Postgraduate Education Committee of the International Psychoanalytic Association and a Fellow of the British Academy. He is a clinical psychologist and a training and supervising analyst in the British Psycho-Analytical Society. His work integrates empirical research with psychoanalytic theory, and his clinical interests center around borderline psychopathology, violence, and early attachment relationships. He has published over 300 chapters and articles and has authored or edited several books. His most recent books include *Psychoanalytic Theories: Perspectives from Developmental Psychopathology* (with M. Target); *What Works for Whom? A Critical Review of Psychotherapy Research* (with A. Roth); *Psychotherapy for Borderline Personality Disorder: Mentalization-Based Treatment* (with Anthony Bateman); *Mentalization-Based Treatment for Borderline Personality Disorder: A Practical Guide* (also with Anthony Bateman); *Reaching the Hard to Reach: Evidence-Based Funding Priorities for Intervention and Research* (with Geoffrey Baruch and David Robins); and *Handbook of Mentalization-Based Treatment* (with Jon Allen).

CHAPTER

5

The Use of SSRIs in the Treatment of Childhood Depression: A Scientific Dialectic

Peter Fonagy

1 Introduction

Major depressive disorder (MDD) is a severe, recurrent problem in children and adolescents affecting approximately 2 percent of individuals under the age of 18.[1] Children with depression experience a range of psychosocial and academic problems and there is an increased risk that they may attempt suicide, or engage in self-harm or substance abuse.[2] A substantial subgroup of young people with MDD have recurrent syndromal and subsyndromal episodes of depression associated with psychosocial impairment and the long-term outcome is poor for a significant proportion of these individuals.[3] Depression is classed as mild, moderate or severe depending on the number of symptoms, which include persistent sadness or low or irritable mood, fatigue, poor or increased sleep and poor concentration.[4] Clinical manifestations in children and adolescents are similar, although prevalence is far higher in adolescence and melancholic symptoms are more prevalent.[5] Depressed children are normally offered multimodal treatment that includes family interventions and psychological therapy on a group or individual basis, in addition to pharmacological agents.[1,6]

This essay is intended to be a contribution to the so-called "antidepressant debate," which gained momentum in 2005 with discussion over the use of selective serotonin reuptake inhibitors (SSRIs) in the treatment of childhood and adolescent depression. This debate has a lengthy and complex history in the context of the pharmacological treatment of major depression in adults.[7–10] It is important to note that this controversy specifically concerns the use of SSRIs for MDD, and not for the treatment of obsessive–compulsive disorder (OCD) or other anxiety-related conditions where its usefulness is largely undisputed .[11,12] This chapter will begin by reviewing some of the criticisms commonly raised in relation to the randomized controlled trial (RCT) literature on the use of SSRIs in the treatment of depression, looking at both the broad context of treatments for depression and at particular instances where SSRIs have been used to treat pediatric groups. We will then interrogate and update these critical assessments with some more recently published findings and reconsider whether the critiques are still valid. Finally, we will consider some of the dilemmas raised for scientists by the work of translating research evidence from clinical trials into recommendations for patients.

2 The thesis

Scientific concerns related to the reporting of clinical trials of psychopharmacological medications run deep and go beyond the relatively small field of pediatric psychopharmacology. There is, and has always been, general unease in the scientific community regarding reliance on industry-sponsored drug research. For example, the outgoing editor of the *British Medical Journal* recently published an article suggesting that a number of medical journals could be seen as extensions of the marketing arm of pharmaceutical companies.[13] He proposed that these journals should cease publishing full descriptions of clinical trials. Journal pages should, he suggested, be restricted to the peer reviewed critical evaluations of studies funded by commercial interests. If industry sponsorship is a controversial issue in general medicine, it should not surprise us that the debate is ferocious in relation to the use of psychoactive medication and that it is particularly intense in relation to the use of antidepressants.[14]

There are three major threads to the arguments against the use of SSRIs in the treatment of major depression. First, it is suggested that the complex and arbitrary nature of psychiatric diagnostic categories compromises the claim that RCTs can be conducted as "medical experiments." Second, it is claimed that the paradigm of RCTs is not sufficiently broad to capture the complex impact of psychoactive drugs. Similarly, it is asserted that the lack of evidence of functional recovery following drug treatments and the wide use of "surrogate" measures of therapeutic outcome means that outcome is assessed on an arbitrary metric.[15,16] Third, it is argued that adverse reactions to SSRIs have been under-reported, are under-recognized and are selectively excluded from reviews of clinical trials to protect vested interests, to the point that the integrity of the research enterprise may be legitimately questioned.

Let us try to gather together the arguments within each of these threads in turn. Most now agree that the clinical entity of MDD is in many respects unsatisfactory. MDD probably represents a heterogeneous group of disorders that are likely to respond differently to specific treatments. Systematic reviews of the treatment evidence, such as the guidelines produced by the UK's National Institute of Clinical Excellence (NICE), underscore the difficulties posed by variation in the responses to the treatment of different individuals.[17] As a consequence, evidence is characterized by frequent partial replication of findings and moderate effect sizes and a diffuse approach in which an unusually large range of treatments have been applied to the disorder.[18] The heterogeneity of the condition is also underscored by etiologic studies, which, for example, show that life events are likely to be associated with depression only in those individuals who have one particular polymorphism of a serotonin receptor gene.[19] The logical constraints of the RCT demand a well-defined clinical group. The high level of comorbidity associated with depression is thought by many to jeopardize the claim of clinical homogeneity.[20] RCTs must be internally valid (i.e., design and conduct must keep the possibility of bias to a minimum) and to be clinically useful the result must be linked to a definable group of patients in a specific clinical setting. The issue raised by critics concerns this latter criterion: that a diagnosis of depression is simply not sufficiently distinct to permit the application of experimental methodology.

Some authors generalize the issue of diagnosis further to issues of etiology.[9] It is claimed that RCTs exploring the treatment of depression using pharmacological agents such as SSRIs address complex cases of emotional suffering that do not readily fit a medical model of illness. A disease model requires a specific pathophysiology and a relatively clear syndromal picture such as, for example, the failure of insulin production in Type I diabetes. This does not amount to a denial of the fact that depression is a physical disorder. Excepting those holding a dualistic position on the body–mind issue, the brain is seen as the mediator of despondency, demoralization, disappointment, discouragement and a sense of defeatedness which are all part of the picture presented in a depressed state. Rather there is controversy over the part played by social as opposed to physical determinants in the condition-in other words, aspects of the etiology that are beyond the scope of the pharmacological RCT.[21]

The second, related argument put forward by those concerned about the use of SSRIs also points to the susceptibility of depression to social influence. The influence of social factors on depression is brought into sharp relief by the powerful placebo responses that the psychopharmacological trials have themselves demonstrated. To expand on this argument, it is assumed that any RCT is a "proof of principle." It ensures that under controlled conditions the causal relationship between a treatment and therapeutic changes may be ascertained. The laboratory model is guaranteed by the so-called "placebo arm" of the trial. However, if depression, or at least some subtypes of it, pertains to the individual's social conditions rather than an endogenous pathogenic brain state, then we might expect that social manipulation such as the administration of a high credibility placebo could produce a very significant change in the condition. This is in fact the case. A highly publicized review of pharmacological studies up to the year 2000 by Kirsch and colleagues[22] demonstrated that at least 82 percent of the treatment response of medicated patients was duplicated in placebo-treated patients. Further, the same review showed that the benefit on continuous scales of treatment response across trials was statistically significant but clinically quite insignificant (1.8 points on a 50 point scale). In bringing this analysis up to date Cohen and Jacobs[9] identified seven RCTs of SSRIs carried out in 2006. None of these showed SSRIs to be superior to placebo and three trials showed placebos to be more effective than active treatments.

Skepticism about efficacy is equally strong in the more sparsely populated field of pediatric trials of SSRIs. Even large-scale trials which have produced good evidence of significant treatment outcomes, such as the multi-site Treatment of Adolescent Depression Study (TADS), have received highly critical appraisal. In a commentary on the TADS trial in the *British Medical Journal*, Jureidini and colleagues[23] wrote: "The small or absent advantages of fluoxetine on other endpoints … shows that fluoxetine, like all other antidepressants, is of doubtful clinical importance for children" (p. 1343). As we shall see, this criticism is unreasonable and was vigorously refuted by the investigators. The comment is included here to indicate the strength of feeling against antidepressant medication even in the medical community.

Of course, with psychoactive drugs, blinded placebo-controlled trials often resemble open trials because the side effects of the drug signal to patients that they are in the active treatment arm. Studies with so-called active placebo might be thought of as controlling for some of these effects by generating side effects.[24,25] Strikingly few studies have contrasted modern antide-pressants with active placebo. This design is not uncommon in other areas of medicine where placebo responses are considered to represent a major confound (e.g., in pain treatment). In the absence of active placebo, ratings by an independent clinician cannot be assumed to be "blind" as the presence of side effects would immediately and inevitably alert an informed rater that the patient has been assigned to the active treatment arm.[26] In sum, the placebogenic nature of antidepressant treatment raises questions about the confidence we can have in the "proof of principle" of effectiveness that RCTs of SSRIs provide.

The third issue highlighted by those concerned about the use of SSRIs is the selective way that drug trials are reported. This is potentially a very broad issue, which touches on the limitations of the knowledge that we may gain from the vast majority of RCTs about clinical practice, since trial samples are rarely representative of the population which is likely to be the recipient of the treatment under investigation [27]. The issue of whether RCTs can be used in a more general context is frequently discussed in the literature on evidence-based medicine[28,29] and questions about how the representative patients are, different levels of severity, treatment context and treatment duration and concurrent treatments (amongst other concerns) will not be covered here. However, comprehensive reporting is crucial if a trial's external validity is to be evaluated.

This aspect of the generalizability issue concerns the SSRI debate very directly. In particular, the extent to which trials may be considered to represent adverse drug reactions realistically has become a central concern. Children and adolescents have had adverse reactions where SSRIs have been used including suicidal ideation[30], growth suppression[31], lethargy and lack of motivation[32], hostility, violence[33], amongst other forms of behavioral toxicity.[34] In the evaluation of any treatment risks have to be assessed against benefits, since no treatment that is not completely inert can be totally without unintended effects. However, the measurement and reporting of side effects is often less thorough and less systematic than the documenting of treatment effects. The high prevalence of polypharmacy in routine clinical practice is of great concern to child psychiatry. Information about the adverse effects of polypharmacy is not even available as individuals on multiple medication are usually excluded from trials.[35]

Evidence from a series of placebo-controlled randomized trials suggested that the prevalence of suicidal ideation had doubled with SSRI use, alerting the FDA and the UK's Committee on Safety of Medicines to the possibility that the risks involved in using SSRIs for depressed children and adolescents might outweigh the benefits.[36,37] The sample set included over 4 500 children and adolescents from 24 studies and the behaviors in question- spontaneous suicidal adverse events (SAEs) encompassed intent to self-harm, actual self-injurious behavior, preparatory action toward imminent suicidal behavior or thoughts about wanting to be dead. Over 90 percent of these events occurred during the treatment phase of the trials; the incidence of SAEs across trials varied from 0 to 8 percent.

These findings contradicted the general belief that psychoactive medications are both safe and effective. Controversy was increased by the apparently selective publication of trial findings. At least one meta-analytic review demonstrated that the published literature painted a more favorable picture of the use of SSRIs in childhood depression than the combination of published and unpublished data might have done.[38] Of course selective publication represents the greatest threat possible to the argument for the generalizability of trial findings. Given that RCTs are an essential step before pharmacology companies can market products, even a mistaken impression of selective reporting would be devastating for the confidence that consumers can have in the validity of the evidence-based guidance available to the well-informed clinician.

Given the constraints of current licensing procedures, RCTs are marketing issues for the manufacturers of pharmacological products. The review that produced evidence opening the possibility of selective reporting of pediatric psychopharmacology trials was met with reactions of great disappointment and consternation.[39–41] Confidence in the integrity of clinical trials was severely undermined.[42] The combination of very real risks and selective reporting generated considerable anxiety and soul-searching, with suggestions of extensive "infiltration" of vested interests into supposedly independent scientific bodies such as Diagnostic and Statistical Manual (DSM) and Food and Drug Administration (FDA) expert advisory panels.[43,44] There was already considerable skepticism about the ability of research organizations such as medical schools to adhere to the standards of integrity, independence and impartiality which the international committee of medical journal editors specified as the minimum standard for trial design, data access and publication rights for industry-sponsored research.[45]

Of course, suspicion that bias may have been introduced into science by industry interests did not stop here. Further wide-ranging concerns were voiced. For example, there were accusations that pharmaceutical companies hired professional writers to ghost-write both trial reports and literature reviews appearing under the name of leading academics.[46] Journalistic and frankly sensationalist interest quickly focused on financial ties between the authors of papers favoring antidepressants and wealthy multinational drug manufacturers.[47,48] The difficulty in obtaining unpublished data on adverse drug events was highlighted by Bennett and colleagues.[49] It seemed that the underreporting of adverse drug reactions, particularly for antidepressants, extended the sense of mistrust to all.

In brief, depression was seen by some as a construct invented by pharmacological companies in order to sell ineffective and dangerous drugs to unsuspecting adolescents.

3 Antithesis

But of course, so far we have only presented one side of the story. The meta-analysis of placebo-controlled trials undertaken by the FDA did indeed identify a twofold increase in the risk of suicidal behavior and suicidal ideation associated with the use of antidepressant medication.[36,37] Pooled data from 4853 participants identified 130 youths with at least one spontaneous SAE, and gave evidence of an unacceptably high risk ratio for any disorder, a figure estimated at 1.95 and in MDD trials alone at 1.66 (CI 1.02–2.68). But careful scrutiny of the evidence can of course yield either more worrying or more benign interpretations of findings.

Emerging data suggested that SAEs were most likely to occur early on in the treatment, a finding consistent with the possibility that SAEs were influenced by the side effects of SSRIs.[50] Alternatively, it is possible that suicidal ideation is particularly high before treatment commences and reduces with increased duration of SSRI treatment.[51,52] Reviews have noted the relatively high frequency of treatment limiting side effects.[53] However, there are major limitations in using combined data from multiple trials of different medications to identify overall risks and benefits for SSRIs as a group. It makes little sense to combine the side effects profile of an effective medication such as fluoxetine with that of venlafaxine, which has very limited effectiveness for this condition. The most common side effect is "behavioral activation" characterized by increased impulsive behaviors, irritability, agitation and silliness in 3–8 percent of patients.[54] However, these are reported in trials where patients were started on quite high doses of the drug. In the light of the worrying side effects, most authorities recommend starting with a low dose and implementing slow dose increases with treatment monitoring for clinical response.[55] In a sense, such debates do not bear on the issue of risk but they do color the interpretation we place on the figures.

More relevant than arguments about the meaning of risk are new data that have clarified various aspects of the debate. Since the original review by Whittington and colleagues, over 1 000 children have been added to the database.[56–59] A new meta-analysis has been reported[12] and the rather depressing scientific picture may be considered to have improved substantially.

It is still not the case however that we can now see SSRIs as the panacea for childhood depression. The Bridge *et al.*, meta-analysis[12] identified an overall response rate to the active treatment arm at 61 percent and those in the placebo arm at 50 percent. The number needed to treat (NNT) before a child or adolescent patient with depression can be assumed to have benefited from SSRIs is large: 10 individuals (95 percent CI: 7, 15). The effect size is 0.25, which by any standard is small. It is even smaller for continuous measures of symptoms of depression: 0.20 (95 percent CI: 0.12, 0.29). The risk of suicide (suicidal ideation and attempts) is still higher in the SSRI-treated group than in the placebo condition (3 versus 2 percent) yielding a number needed to harm of 112. Some clinically obvious, yet important moderators to emerge from this meta-analysis included the smaller effect size for those with longer duration of depressive illness and those treated in larger multi-site studies. Not surprisingly, children showed larger placebo effects than adolescents. A more surprising absence of findings, given the nature of the controversy, might be that the source of funding for the studies was not examined in relation to outcome. Given that all federally funded studies were of fluoxetine, the source of funding variable would have yielded interesting variance.

Even in this more recent analysis, the observation that an increased risk of suicidal ideation is associated with SSRIs appears to be robust. However, there are no completed suicides. None of the studies were specifically designed as trials of the impact of SSRIs on suicidal ideation. Even more importantly in the current context, there were very substantial differences between SSRIs in terms of effectiveness (see Figure 5.11). The figure shows that most of the SSRIs tested do not deserve consideration as effective treatments and that the meta-analysis is based on significant heterogeneity of findings. Only fluoxetine appears consistently to outperform placebo. Under these circumstances, looking at overall levels of risk makes little sense, and

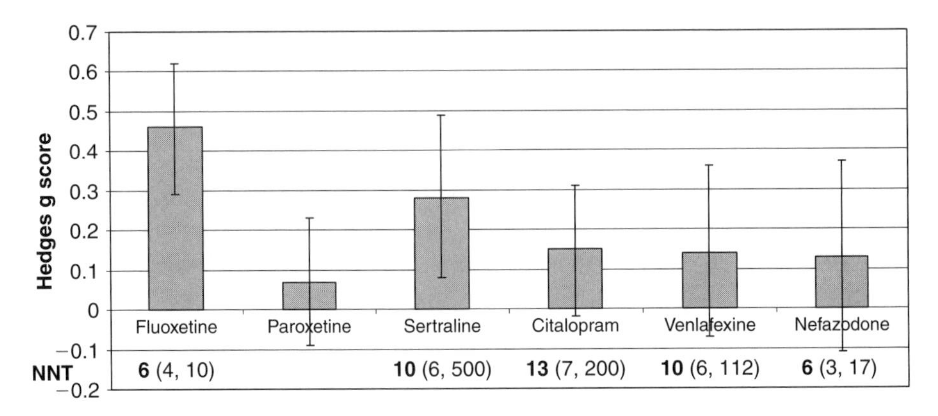

FIGURE 5.11 Efficacy outcomes for antidepressant-treated and placebo-treated depressed groups (Bridges *et al.*, 2007 *Journal of American Medical Association*).

we need to look at the most informative individual studies to understand the risk-benefit profile of this treatment.

More detailed analysis of two trials that recently reported on the performance of fluoxetine offers important clinical guidance. The TADS trial explicitly explored the effect of fluoxetine relative to placebo in cases stratified for severity over a 12-week period.[50,58,60,61] Both the TADS trial and the British RCT of fluoxetine, the ADAPT trial[59], suggest that antidepressants are useful for adolescents suffering from moderate to severe depression. In the TADS trial, 12-week treatment outcomes for a combination of fluoxetine and Cognitive-Behavioural Therapy (CBT) were significantly better than for fluoxetine alone. At 12 weeks CBT was no better than placebo. The effect size for combination treatments relative to placebo at 12 weeks was .98; for fluoxetine, it was .68; for CBT, it was −.03. The NNT for combination treatment was 3; for fluoxetine alone it was 4 and 12 for CBT. The addition of CBT increased response rates by 10 percent relative to fluoxetine, but fluoxetine had a substantially higher response rate than CBT for placebo. The rate of adverse events was consistent with previous studies of fluoxetine. Twelve percent of individuals treated with fluoxetine reported clinician-identified suicidal events compared with about five percent of those on placebo or CBT treatment. The combination of fluoxetine and CBT indicated a significant benefit relative to placebo on 15 out of 16 endpoints. The addition of fluoxetine showed superiority over CBT on 14 out of 16 endpoints.

The report on the long-term effects of the TADS treatments (at 36 weeks) no longer includes a placebo group, as the trial was unblinded. However, the outcomes picture changed substantially. Almost all the samples showed a treatment response: 86 percent of those in combination therapy and 81 percent of those receiving fluoxetine compared to 80 percent of those receiving CBT. Observed cases and intent to treat analysis showed that both combination therapy and fluoxetine alone were superior to CBT alone in bringing about improvement faster by 6–18 weeks but the groups ended up at the same point by 24–36 weeks. Evidence on suicidal ideation and events favored the psychotherapy group. By week 12 patients treated with fluoxetine continued to show more clinically significant suicidal ideation than those treated with CBT. At week 36, 14 percent of those on fluoxetine alone reported clinically significant suicidal ideation as distinct from just 2.5 percent of those in combination treatment and 4 percent of those in CBT. The findings suggested that while CBT added little by way of greater treatment effectiveness, it served to reduce adverse events such as SAEs (at 36 weeks 15 percent for the fluoxetine sample, 8 percent for the combination therapy and 6 percent for CBT alone). In the fluoxetine alone group, there were approximately double the number of emergent suicidal events observed in patients treated with CBT or combination therapy. Notwithstanding the controversial nature of SSRI treatments, the effects of the TADS study confirm the benefits of including fluoxetine in the short-term treatment of adolescent major depression. Fluoxetine accelerates treatment response relative to both placebo and CBT. The TADS trial confirms fluoxetine as an effective treatment of moderate and severe MDD. CBT turns out to be a very reasonable alternative within a few months and seems to mitigate suicidality associated with fluoxetine, perhaps by minimizing the likelihood that ideation will lead to action and improving the individual's capacity to manage stressful events, family conflicts and negative affect.

A similar trial confirming some but not all the findings of TADS was carried out in the UK with what appears to be a more severely depressed sample of adolescents [59]. While strongly confirming the benefits of fluoxetine, at 12 weeks 47 percent of those on fluoxetine and 43 percent of those on combination treatment (CBT plus fluoxetine) were much improved or very much improved. By 28 weeks, these figures increased to 51 and 53 percent, respectively. Symptoms of suicidality reduced over the period of treatment, as would be expected, but in the absence of a placebo-controlled condition we cannot determine whether SSRIs caused an unexpected increase. The majority of patients reported side effects from SSRIs but there was no indication that CBT mitigated suicidality, self-harm or other adverse events.

With children the effects of fluoxetine are likely to be exaggerated by the powerful placebo response that younger children show. The meta-analysis in relation to chronicity suggested that early treatment may be more effective than later intervention using antidepressants. While this finding may be an artifact of severity or treatment response in general, it is important to bear in mind that current advice [1] based on previous meta-analyses is for a very conservative treatment protocol recommending a delay in introducing antidepressants. This may no longer be evidence-based advice.[62] However, notwithstanding intense criticisms of the advice concerning the use of psychosocial treatments as first line treatments for most mild to moderate depression in the NICE guidelines, the use of CBT and other psychosocial therapies remains well supported by evidence from more recent meta-analyses.[63,64]

4 Synthesis

So how can we reconcile the worrying statistics following the FDA investigations leading to the black boxing of SSRIs and the far more reassuring findings of benefits relative to risks reviewed in the last section of this chapter? The difference between the Bridge meta-analysis and the FDA meta-analysis of suicidal risk is largely one of statistical approach. The approach adopted by the FDA assumed that the effects are common across all studies. The approach in the more recent meta-analysis is more conservative, appropriately allowing for heterogeneity between trials. This argument is also made by those who have re-analyzed the FDA meta-analysis using Bayesian statistical methods.[65] The reader who is skeptical about statistical procedures may well express reservations about the idea that statistical methodology should govern whether we advocate for or counsel against the use of SSRIs in the treatment of pediatric depression. However, the practical implications may not be so different. Both statistical models are consistent with the assumption that baseline characteristics such as irritability and previous suicidal ideation predict the iatrogenic response to SSRIs in this group.[66,67] This suggests that attention should focus not on "macro" initiatives such as discouraging the use of SSRIs but rather on developing efficient methods for monitoring the response of children and adolescents to careful trials of antidepressants.

In brief, it seems that in the case of the risk of suicide associated with treatment with fluoxetine, "science worked." Initial findings alerted researchers to the possibility of significant harm. Additional studies and further analyses of existing data demonstrated that (a) the problem was real; (b) it was probably most marked in connection with some relatively ineffective SSRIs (e.g., venlafaxine); (c) the benefit to risk ratio favored fluoxetine and (d) certain psychosocial interventions can be helpful in reducing or eliminating the risk of suicidality associated with the use of the medication, at least in cases of moderate depression. The evidence-based approach turned out to be helpful in reinforcing certainty. It triumphed over intuition and the unsystematic application of expertise or clinical experience by strengthening our grip on cause–effect relationships. RCTs and the integration of evidence from multiple sources in systematic reviews and meta-analyses, alongside the use of integrated indicators of the magnitude of effects of beneficial and noxious factors, allow greater precision in answering practice-oriented questions about the risks and benefits associated with SSRIs. However, as Goldenberg[68] pointed out, whilst evidence-based approaches undoubtedly improve medical practice, the term evidence based should not be understood to be synonymous with best practice in all relevant respects. The story of this particular controversy may be a case in point.

Looking back at the controversy, the literal interpretation of the evidence led the FDA, NICE and the European Commission to issue advice to limit the use of SSRIs in treating childhood and adolescent depression. This clashed strongly with the intuition of many clinicians and generated extensive concern in the field.[62] As Archie Cochrane himself pointed out, "between measurement based on RCTs and benefit … in the community there is a gulf which has been underestimated."[69] Whilst RCTs and systematic reviews represent the most reliable method of determining moderate treatment effects, to be clinically useful the result must also be relevant to a definable group of patients in a particular clinical setting. Looking back, can we see a way in which the moral panic concerning the use of SSRIs could have been avoided?

Three issues seem clear with hindsight. The first concerns the difference between evidence-based treatment and evidence-based practice. Whilst evidence-based treatments are those which have been demonstrated to be effective by a certain minimum number of well designed and conducted RCTs (say two or more), evidence-based practice is the integration of the best available research evidence with clinical expertise in the context of patient characteristics, culture and preferences. Evidence-based practice articulates a decision-making process for integrating multiple streams of research evidence, including but not limited to RCTs.

Second, the more flexible approach of evidence-based practice would include the exploration of correlational data such as observations concerning the apparent impact of the macro initiatives undertaken by national and international bodies in relation to the pediatric use of SSRIs for depression. Epidemiological investigations have highlighted how reducing access to SSRIs has served to increase suicidality in both the US and Europe following the crisis of confidence in drug trials.[55,70] Looking back we see that the profile of benefit and risk for SSRIs changed rapidly as new evidence accumulated. Perhaps there are no obvious desirable steps that could be taken to slow implementation of emerging evidence; any such steps would probably be both unethical and undesirable. However, policy makers must recognize that the side effects of dramatic shifts in policy may be more severe than the effects these intend to prevent, and further, that there may be wisdom in resisting the impulse, wherever possible, to constrain the clinician's freedom to make responsible and informed pragmatic decisions about individual patient care.

Third, the problem concerning SSRIs arose out of the close association between industry and science. In most Western countries scientists are exhorted by their governments to pursue partnerships with industry, in order to increase the direct applicability of scientific endeavor as well as reduce taxpayer costs by commercial subsidy. Of course such initiatives run considerable risks and the close association of trialists with the pharmacology industry may be an area that warrants particularly close scrutiny. Interestingly, a number of industry-funded projects unpublished at the time of the Lancet review have since appeared. For example, Emslie and colleagues[71] have reported (with a little delay – the study was carried out 7 years before publication) paroxetine to be ineffective in the treatment of childhood and adolescent depression. However, industry sponsorship is selective. Certain types of studies are not going to be funded by manufacturers of medicines. In the specific case of childhood depression, the literature is still lacking a trial designed to examine the risks associated with the use of fluoxetine. Such a prospective study would require a sample size large enough to detect the possibility of a 1:100 event such as an SAE or other similar low-frequency adverse reactions.

Until the results of this hypothetical trial become available we should feel comforted that many of the initial doubts expressed about the use of SSRIs for children have been addressed by additional research, and that both the initial warning siren and the more recent sounding of a temporary and selective "all clear," have been achieved by solid, focused scientific inquiry.

David H. Smith, Ph.D.
Director, Interdisciplinary Center for Bioethics
Yale University
New Haven, CT, USA

David Smith joined the faculty of the Department of Religious Studies at Indiana University Bloomington in 1967. While at Indiana he served as Chair (1976–1984) of the Department of Religious Studies and from 1983–2003 as Director of the Poynter Center for the Study of Ethics and American Institutions. Smith is the author most recently of *Partnership with the Dying: How Medicine and Ministry Should Work Together* and editor of *Doing Good: Moral Opportunities and Pitfalls*. His article "Stuck in the Middle" appeared in *The Hastings Center Report* in January 2006.

Since his retirement from Indiana University Smith has served as Visiting Professor of Bioethics at Yale University (2003–2004, 2006–2007). In 2004–2006 he was Fredrick Distinguished Visiting Professor of Ethics at DePauw University, where he taught courses in biomedical ethics, the morality of killing, and moral issues in doing good. Smith currently serves as the Director of the Yale Interdisciplinary Center for Bioethics.

Aaron Klink, M.Div.
Westbrook Scholar of Theology and Medicine
Duke University School of Divinity
Durham, NC, USA

Aaron received his B.A., in Religion and Political Science from Emory University and his Master of Divinity degree from Yale Divinity School. He was a chaplain resident at Yale New Haven Hospital and an intern at the Yale Interdisciplinary Bioethics Center, where he now serves on the Advisory Committee of the End of Life Issues Working Group. Aaron has done pastoral work in Lutheran congregations in New York, Pennsylvania and Connecticut. He is currently Westbrook Scholar in the Program in Theology and Medicine at the Duke University School of Divinity.

Julius Landwirth, M.D., J.D.
Associate Director, Interdisciplinary Center for Bioethics
Yale University
New Haven, CT, USA

Dr. Landwirth began his professional career in 1963 as a practicing pediatrician in Orange, Connecticut. His interest in teaching and research led him to serve as the first full-time Director of Pediatrics at Bridgeport Hospital and a faculty appointment as Clinical Associate Professor of Pediatrics at Yale. After a decade during which he developed a Yale-affiliated pediatric training program and the first seminar series in medical ethics, he moved on to become Director of Pediatrics at Hartford Hospital and Professor of Pediatrics at University of Connecticut Health Center. There he founded the hospital's first ethics committee which he chaired for 14 years. During that time he also served as Medical Director of Newington Children's Hospital (the forerunner of the current Connecticut Children's Medical Center). He also served as consultant in health care ethics for the Connecticut Medical Society.

After completing his law degree in 1987, at the University of Connecticut, and following his retirement from hospital positions in 1996, he pursued his interest in international health and ethics by joining the Connecticut-based Albert Schweitzer Institute as its Director of Health Care Programs. In a 6-year collaboration with the Open Society Institute he organized educational programs in 15 countries throughout Central and Eastern Europe and Central Asia, including Mongolia.

In his current capacity, as the Associate Director of the Yale University Interdisciplinary Center for Bioethics, he also serves as Associate Director of the Donaghue Initiative in Biomedical and Behavioral Research Ethics and has been appointed to the Connecticut Stem Cell Advisory Committee and the Committee on Ethical, Legal and Ethical Issues in Genomics.

CHAPTER

6

Media Coverage of Stem Cell Research

David H. Smith, Aaron Klink and Julius Landwirth

Soon after the isolation of human embryonic stem cells, speculation began about their possible uses for research and treatment related to neurological problems such as Parkinsonism and Alzheimer's disease. In this paper we will discuss two recent episodes of press coverage of uses of stem cells as they do, or may, relate to neurology. Predictably any moral verdict on the actions or motives of key players will have to be circumspect, but some morals of the stories will emerge pretty clearly. We will then turn to the professional responsibilities of scientists and journalists, giving some special attention to the role of press releases – documents that arise in a context that belongs to neither guild.

1 Scientists, journalists and the practice of *journalistic* framing

We begin with some general comments about the communities of scientific researchers and science journalists.

Studies of the complex interactions between journalists and scientists are not lacking. To take just one important example, in the 1980s Dorothy Nelkin carefully analyzed their arguments and frustrations with each other. Nelkin stressed the differences between the two communities and those differences are surely important, as our paper will show.[1] However, it is important at the outset to see that the disagreements may be part of a family feud. Both communities are focused on getting and communicating as accurate or true an account of states of affairs as they can. Both are indifferent to motive, intolerant of baloney and committed to tough mindedness. Anyone who has worked with diverse professional groups will have been struck by these parallels.

Someone may say that scientists are interested in discovering new truths, whereas journalists only report what others have discovered. But in fact the tradition of investigative reporting is a proud part of the journalistic community, even if – as Nelkin and others have pointed out – it is more commonly applied to political than scientific news. Or it may be claimed that the scientist is only secondarily concerned with communicating what she has learned; communication is of the essence of journalistic activity. This gets at something right, for the idea of a scientist who can't write isn't as problematic as of a journalist whose style is awful. But much important journalistic writing is pretty marginal, and science is a social enterprise requiring accurate and intelligible communication of the results of one's work to – at least – a community of one's peers.

Moreover, the commitment to honesty and accuracy is part of the professional credo of both groups. Eradication of fabrication, falsification and plagiarism – the core of much of the recent stress on professional integrity in science – could easily be treated as a journalist's credo. Making up a story, even in a good cause, is the unforgivable sin for both groups.

Why then is getting journalists and scientists together something like herding cats? At least part of the answer lies in the issue of framing, as Nisbit and colleagues have pointed out in an important article.[2]

Both a scientific paper and an article in a more general circulation publication must be framed or shaped. Neither is simply a straightforward report of what happened. The scientific paper is a work of art, following an accepted stylistic form, and designed effectively to communicate to peers the core replicable results and the processes followed to lead to them. Difficulties encountered along the way, the role of dumb luck, reasons for publishing this portion of the ongoing research at this time – none of these things has an essential role in the paper.

Journalistic framing follows a different logic. The journalist wants to fit the scientific event into a larger story, ideally in at least two senses. First, to assure space in the paper and a readership of what does appear, a journalist wants to hook the new event or discovery into a larger story that has ongoing public interest such as cures for cancer, heart disease, or, in our case, Parkinsonism or Alzheimer's disease. That is the context in which the story is of interest to a larger public that is relatively unconcerned with the incremental progress – three steps forward and two steps back – of science. Sometimes a story may be of minimal interest in this larger sense but can be made more gripping by setting the new event in the context of the story of the work of a researcher, laboratory, company or university. That is, the event may be of general interest because it was done at Johns Hopkins, or Harvard, or because it was done in Korea – and not in the United States. It may be important that the work was done by a venture capital firm.

A journalist, therefore, might be particularly concerned with some issues of motivation, context and competition, problematical relations within a laboratory or company, and funding difficulties that seem irrelevant to a scientist. Why? Because no journalist, certainly no science journalist, can simply assume her story will be used, let alone read. Newspapers are in the business of making money; editors and publishers refer to the "news hole" which is the amount of space in any edition that can be devoted to news as opposed to commercial advertising, classified adds, etc. The fact of competition for space within the paper is no justification for sloppy reporting; it goes some way to explaining the constraints on a science journalist.

The days in which scientists were able to ignore the press are long gone. For the sake of the public good as well as for securing funding and reputation scientists want effective and positive press coverage. Indeed, as Nisbit and colleagues write sources and journalists are in dynamic interaction producing a "negotiation of newsworthiness and a cooperative manufacture of news." We live in a "mediated democracy" in which events are not only covered by the media, but they are shaped by the media and their sources. In some very important senses the event is what the media say it is.

One final point before we turn to our two stories. Sources for media accounts may include scientists, institutions, sponsors, clergy and – more recently – bioethicists. The relative effectiveness and clarity of these sources will shape the way a given story about stem cell research is framed. For example, stem cell research stories may cast the "news" primarily in relation to the scientific background, a salient moral issue of general interest such as cloning or research on embryos and fetal tissue, or the political strategy of the Bush administration, or patents, public opinion. These questions of framing are of great importance.

2 Some recent work by Advanced Cell Technology

"Human Embryonic Stem Cell Lines Derived from Single Blastomere" was an Advanced Online Publication in *Nature* in August 2006.[3] Stem cells are important because they are pluripotent – given the right stimuli and growth media they can become any cell of the body. Thus they are potentially invaluable to research and therapy, including neurological research and therapy related to spinal cord trauma, Parkinsonism or Alzheimer's disease.

Stem cells can be found in many tissues of the body and in umbilical cords, but the cells of the greatest scientific utility come from human embryos. So far those cells have been removed from the inner cell mass of embryos at about 5 days gestational age. The removed cells are then cultured and reproduce to form a "stem cell line." The result is great for research, but the procedure entails the destruction of the embryo, and thus raises again the moral issue that has created the most public controversy since the isolation of human embryonic stem cells in the late 1990s.

The *Nature* paper by Klimanskaya *et al.* suggested a possible way around the controversy arising from destruction of the embryo: developing stem cell lines from even earlier human cells called blastomeres. Preimplantation genetic diagnosis (PGD) – which amounts to a genetic test on an embryo to be implanted in an *in vitro* fertilization (IVF) procedure – involves taking one blastomere from the earliest embryo. Why not use such an extracted cell to develop stem cells? As in PGD, the remaining cells are highly likely to continue to grow and develop into a complete organism. Thus the embryo runs a slight risk, but it is not destroyed and the controversy is bypassed. Or so it might seem.

The scientific question for the Klimanskaya paper was whether stem cell lines could be developed from a blastomere extracted from a human blastomere. An Advanced Cell Technology (ACT) paper from a few months earlier had shown that the procedure would work with mouse embryos. And the Klimanskaya paper showed that the answer was "yes" for human embryos as well – in the particular circumstances of this experiment. Careful readers of the paper could see that the donor embryos were used up or destroyed in the procedure and that only two cell lines were established from the ninety-one cells the researchers started with. The paper appeared in advanced online publication.

ACT's original news release began by saying that the blastomere derived stem cells came from "an approach that does not harm embryos" a claim that is correct about the PGD technique but not about the research on which ACT was reporting.[4] Indeed Robert Lanza the senior (but not first) author on the paper was quoted in the release as saying "We have demonstrated, for the first time, that human embryonic stem cells can be generated without interfering with the embryo's potential for life..." (On June 21, 2007 Lanza claimed to the International Society for Stem Cell Research, meeting in Australia, that ACT had "successfully produced a human stem cell line without destroying an embryo."[5] No scientific publication is cited in ACT's press release on his 2007 statement.)

Criticism of the ACT claims was quickly published. On September 6, 2006 *Nature*'s web site noted that German newspapers claimed Lanza had "hyped" his results.[6] The ACT paper authors acknowledged that "none of the biopsied embryos were allowed to develop in culture." They recognized that not all blastomere cells might be capable of forming embryonic stem cells, and they added more information on the handling of embryos to the already posted publication. This constituted an "Addendum." The authors also published a "Corrigendum" that said the last sentence of the article should read "Notably, individual morula (8–16 cell)-stage blastomeres have not been shown to have the intrinsic capacity to generate a complete organism in most mammalian species."[7]

Immediately after the August 24 online publication the *Los Angeles Times* ran a story that ACT picked up for its web site.[8] The *Times* story referred to a technique that "does not require the destruction of embryos" and framed the science as part of the political and moral debate over the use of stem cells derived from destroyed embryos. A White House spokeswoman and Richard Doerflinger of the US Conference of Catholic Bishops were cited on that controversy. The risk of the technique to the developing embryo was assessed by sources outside ACT and pro-life lobbies. Ronald M. Green, chair of ACT's Ethics Advisory Board argued that the new procedure was unusual because it was a scientific breakthrough that helped "resolve" an ethical problem.

The *LA Times* story on the whole is quite good and balanced, especially when we remember that it was produced right after the announcement. Its framing is understandable and clear. But even this story allows Doerflinger's claim that PGD has led to long term effects on persons conceived through IVF to go unchallenged, and it does not give much attention the difference between the ACT claims and the details of the original paper.

The New York Times story by Nicholas Wade on August 23, 2006 has a less apparent political and moral framing, but still focuses on those aspects of the discovery.[9] Wade reported Ronald Green's contention that the blastomere derivation technique might enable freeing of federal funds for embryonic stem cell research; he followed that up by contacting the head of a stem cell task force at National Institutes of Health (NIH) who admitted uncertainty on the point but who added that children who started as embryos that had PGD testing "seem to grow into babies as healthy as other babies born by IVF." This disagreement with Doerflinger's statement about risks to future children, may explain the fact that Doerflinger used another argument in his conversation with Wade.

Wade also contacted Leon Kass, first Chairman of the President's Bioethics Council, who noted that the long term effect of PGD testing on embryos is simply unknown and was skeptical about the likelihood of the discovery resolving the ethical problems of embryo use. Wade did more checking with sources outside the political controversy than the *LA Times*, but even he gives little attention to the difference between technical claims and press release.

Rick Weiss in the *Washington Post* provided what was probably the best coverage of that problem.[10] He quotes ACT Ethics Advisory Board Chair Green as saying "You can honestly say this cell line is from an embryo that was in no way harmed or destroyed." Followed shortly by the factual statement that "The researchers took not just one cell from each [embryo] but as many as they could get – destroying some of the embryos..." Doerflinger is cited but much more briefly; the political context is fully treated, but the science gets the best explanation of any of these three stories.

Predictably coverage in the press for scientists was thorough and accurate. Writing in *Nature* at the time of the hard copy publication of the ACT paper, Joe Leigh Simpson offered a full, helpful explanation of what had been done and what it might enable.[11] Simpson clearly sketched both the limits and the prospects of the research to date, gave important attention to the moral issues at stake, then concluded that the research "is realistic and offers an attractive way out of an ethical conundrum." Even those who may disagree with that claim in ethics would have little to disagree about in the article, for it offers clear data on which a differing verdict might be based.

Coverage in *The Scientist* was particularly strong.[12] Writing more than 6 months after the publication of the original paper Charles Q. Choi did not rehearse the controversy over the difference between science and press release; he did quote several impartial sources as saying that the replicability of the science had yet to be shown, that the technique as so far demonstrated was inefficient, and generally that the paper was more suggestive than conclusive. *Blog* comments on the story, however, included scathing comment on the ACT publication strategy as well as a couple of endorsements from parents and grandparents who had seen wonderful results from stem cell therapies.

Before passing from reports on this discovery to consideration of another situation, there is a further aspect to the ACT press release policy that we want to mention. It concerns "forward looking statements." "Forward looking statements" is a term of art from the world of corporate finance; it refers to statements about future hopes and expectations by corporate management. The rough idea is that corporations cannot be held liable for expected or hoped for results that do not materialize after having been announced in "forward looking statements." The ACT disclaimer concludes with this one sentence paragraph:

> Forward looking statements are based on the beliefs, opinions and expectations of the company's management at the time they are made, and the company does not assume any obligation to update its forward-looking statements if those beliefs, opinions, expectations, or other circumstances should change.[13]

This may allow the company to retain misleading information, originally published in good faith, in the public domain. We will return to this practice in our conclusion.

3 Neuron replacement in Hopkins rats

Douglas Kerr and associates at Johns Hopkins University reported research on stem cell transplantation into rats in the *Annals of Neurology* in June 2006. They wrote that

> [S]pinal motor neurons can be generated officially by exposing mouse embryonic stem cells to retinoic acid and developmental morphogen Sonic Hedgehog chemical agents or chemical agonists of Shh. In this

> paradigm, RA serves to neutralize and to establish a caudal positional identity for the ES cells, whereas Shh produces ventral positional identity. In response, many ES cell imitate a motor neuron specific transcriptional pattern, and acquire immunohistorical and electrophysiological features of mature motor neurons.[14]

Compare this with an Associated Press report "Scientists have used stem cells and a potion of nerve friendly chemicals not to just bridge a damaged spinal cord but actually re-grow the circuitry needed to move a muscle, helping partially paralyzed rats walk."[15] The treatment was referred to as a "cocktail."

The *WebMD* web site comes somewhere in the middle quoting in the first paragraph the modest claim of the Kerr team's *Annals of Neurology* report that the method "may be a potential therapeutic intervention for humans with paralysis." WebMD doesn't mention the steps necessary to transform Kerr's experiments with mice into treatments suitable for human beings.[16]

Some sources appeared over and over again in several different publications. Among them was Douglass Kerr's comment that he hoped the results [lead to] "the appropriate tempering of expectations of stem cells." Another was a vivid metaphor for stem cell treatments when he recounted that some of his patients think, "Oh, I'm going to pull into the stem-cell station and get my infusions of stem cells, it's never going to be like that." It is interesting to note how careful and measured Kerr's statements about his own research were. For instance, he only claimed that the results, "hinted at the promise" of stem cell research, but did not claim he had found the solution to all problems. It was only when the press "hyped" or "spun" the article that stronger claims were suggested.

What is this research applicable to curing? The most expansive claim was made by *The Johns Hopkins University Gazette* that titled its article, "Embryonic Stem Cells Awaken Latent Nerve Repair." Despite a modest title the paper enthusiastically claimed that Kerr's "approach could one day repair nerve damage from such diseases as Lou Gehrig's, multiple sclerosis, and transverse myelitis or from traumatic spinal cord injury." They continued that with small adaptations; "the approach may also apply to patients with Parkinson's or Huntington's disease."[17] They are the only source to mention that Kerr's research was in part funded by Families of SMA, Andrew's Buddies (an SMA association) the Muscular Dystrophy Association and it's partner foundation, Wings over Wall Street and last but not least, by the NIH. So the *Gazette* stresses the potential of the research as well as the public's investment in it.

To its credit the *Gazette* was the only publication to clearly point out that the rats were not "totally cured," and only had partial movement restored. (CBS news coverage, which included an interview with Douglass Kerr, also noted that the rats had "restored movement" but were not "totally cured.")[18] Only the *Gazette* was specific about what happened to the rats, namely that they, "gained weight, were more mobile in their cages and measures of muscle strength increased."[19]

Speculation about future applications of the Hopkins research has continued. For instance, the *Mental Health Law Weekly* noted, "The study suggests that similar techniques may be useful for treating spinal cord injury, transverse myelitis, amyotrophic lateral sclerosis, and muscular dystrophy" but did not speak about the intervening tests on larger mammals that needed to be done before it could be used in humans, making it seem like the application of Kerr's discovery was not far away.[20] The Associated Press' *Worldstream* release was more realistic noting in the second sentence of its article that, "Years of additional research [are] needed before such an experiment could be attempted in people."[21] The National Institute for Health was a bit more guarded acknowledging Kerr's breakthrough but also cautioning that the research Kerr's team did had a long way to go before becoming applicable in humans. Elias Zerhouni was quoted in the NIH release as saying, "The successful demonstration of functional restoration is proof of the principle and an important step forward. We must remember, however, that we still have a great distance to go."[22] The NIH press release also was one of the few articles that mentioned the tests had to be done in larger animals, and that scientists were uncertain if the technique would work over longer distances in larger animals. In the same way, the NIH was the only article of that small subset that reports that it will be difficult to test in large mammals because there are no known larger mammals that suffer musical degeneration.

So too, a few newspapers noted the impact of federal disputes over stem cell research, most notably the regulations prohibiting the government to fund research on stem cell lines not in existence prior to a President Bush's Executive Order. *The Baltimore Sun* quoted Kerr as saying, "The amount of money required to take this to clinical trials is ultimately hundreds of millions of dollars … without federal funding that money is going to be difficult to find." *MEDPAGE Today* begins its article with the political frame noting that "The proof of concept study is sure to fan the flames of stem cell debate."[23]

4 So what for ethics?

Some comparisons of these brief scenarios come quickly to mind. Even taking into account the difference in elapsed time between scientific discovery and our analysis, it seems clear that the ACT discovery attracted more coverage than the stem cell neuron transplant. We find this noteworthy as the neuron transplant is considerably closer to the development of a possible therapeutic breakthrough than the ACT research. We hypothesize that the reason for this difference in coverage lies in the fact that the use of embryonic cells has become part of our larger culture's public debate and the ACT story more clearly has a place in that larger narrative. It has a better news hook than the Hopkins story, so it gets more coverage.

The greater coverage may also relate to the fact that from the first ACT press release large claims were made for the possibilities of the use of blastomeres as sources of embryonic stem cells. For whatever reason, perhaps because of its origin in a private research and development firm, ACT press contacts took an assertive and optimistic stance. The contrast with Kerr's moderation is striking, and even the *Hopkins Gazette*, that made larger claims than Kerr himself, was very specific about all the difficulties still to be overcome before therapy on humans could even be tested. The quest for funding – whether in the form of grants, private donations, public funding or investment – obviously complements a desire for fame and status in driving whatever exaggerations occur.

The drive for funding has a great influence on individual scientists, universities and private firms. Federal funding is declining. The American Association for the Advancement of Science reports that funding for the NIH, the organization responsible for funding much of America's basic science research remained flat in 2007. Few private investors want to invest in the most basic scientific research. Patient interest groups often run by those who suffer illness or have lost loved ones to it, want speedy cures not promises that may be too late for themselves or those they love. Knowledge in itself is not always or immediately profitable and for every path-breaking (and profitable) discovery there are a dozen dead-end roads. Therefore, in order to maintain strong research programs, it is imperative for scientists and the institutions in which they work to show their work in the best possible light.

5 Scientific press releases and the media

The important role that the media play in translating medical research into news and influencing public perceptions is well recognized.[24] Therefore it is important to understand how journalists choose the stories they will cover and how they present the information. As mentioned earlier, journalists prefer to frame stories in terms of dramatic narratives of general public interest. It has been pointed out that once an issue is framed by the media, it can be difficult for policymakers or other interests to shift the image of the issue to another perspective.[25] In one study of the news coverage of two major British scientific journals, journalists were more likely to report articles dealing with fatal or rare but interesting diseases, those with sexual implications, those describing new or improved treatments and those involving highly controversial subject matters. Interviewed journalists stressed that "medically worthy information is not necessarily newsworthy."[26]

Since journalists are rarely in the position critically to review the scientific validity and import of the results of newly reported research, they are heavily reliant on sources that describe the substance and broader implications of the work. Prominent among these sources are press releases, which may be issued by representatives of government, industry, religious organizations, interest groups, academic institutions or scientific journals. In their effort to attract the interest of journalists, these sources may emphasize what they perceive to be a key aspect of the report, in that way contributing to the way the story is ultimately framed. Internet postings, on sites such as the Eurekalert (http://www.eurekalert.org), produced by the American Association for the Advancement of Science, offer journalists convenient entry points to press releases from a variety of sources.

The effectiveness of press releases in the chain of communication from scientist to the public is clear. One study found that 84 percent of journal articles referred to in newspaper stories were the subject of press releases by the journal itself.[27] Commentators have raised issues of professional standards and accountability for unbalanced information released by institutional press officers. One nationally prominent science writer has written "Scientists rarely make exaggerated claims when reporting their results in the scientific literature because it is poor etiquette and likely to provoke the scorn of their peers. But news releases are a different matter."[28] Responsibility for distorted reporting in press releases is less clear than the responsibility of editors for the validity of journal articles.

A few published studies have looked at the dynamics of press releases from scientific journals, inquiring about how articles are chosen for press releases, who wrote them, whether guidelines exist, to what extent authors or editors are involved, and whether funding sources are described. One such study found that press releases from scientific journals frequently presented data in exaggerated form, failed to highlight study limitations or financial conflicts of interest and had variable input from editors and authors.[29] Thus, tracing newspaper stories about a scientific advances to their sources in the workrooms of journal press officers may shed some light on the selection, framing and content of newspaper stories.

Press releases from commercial bioscience industry have a somewhat different and very interesting dynamic. For-profit corporations, including bioscience and pharmaceutical companies, have a primary interest in putting out press releases and other forms of communication that may be of interest to potential investors. These releases may also serve a public information function when they become the subject of newspaper articles. Security and Exchange Commission rules are intended to protect the public from false or misleading statements in such releases. At the same time corporations seek protection from litigation from individuals who claim damages resulting from their reliance on incorrect information in a press release about the future potential of the reported scientific advance. Such releases are referred to as "forward looking statements."

In an attempt to strike a balance, Congress passed the Private Securities Litigation Reform Act of 1995 (http://www.lectlaw.com/files/stf04.htm).

The purpose of the Act is to provide investors with valuable information, while protecting them from abuse, as well as protecting the company from frivolous lawsuits. The Act creates a safe harbor for liability from forward looking statements that fail accurately to predict the future.[30,31] Corporations routinely include the term "forward looking statement" in their posted releases. This legal term of art connotes considerable latitude in claims made that, when picked up by journalists and worked into newspaper stories, might be judged by media critics to be exaggerated reporting (hype).

5.1 What's wrong with hype?

Why is it important to resist this pressure and to work toward more careful and balanced coverage? Is it acceptable to continue on a path with considerable hype and more than one misleading statement? In conclusion we offer three kinds of reasons that these forms of reporting should be matters of ethical concern.

First, inaccurate and overly optimistic statements have serious consequences for society as a whole. When the inaccuracies are revealed they discredit science, and science is an enterprise about which many people are already ambivalent. To say that this discrediting is bad for science is a great understatement, for scientific research is dependent on the public for funding – whether the funding comes through government or private sources. Just in terms of self-interest restraint is wise.

But there are other consequences of great importance. For example, seriously ill persons may make decisions about their treatment, employment, retirement or life style on their assessment of the likelihood of the success of some specific breakthrough in Alzheimer's disease or Parkinsonism. Indeed, many physicians may unintentionally mislead their patients on the basis of articles in the popular or scientific press. Furthermore, public policy about embryonic stem research or other aspects of neuroscience depends on well-informed bureaucrats, legislators, administrators and citizens. Responsible and progressive government must be based on an accurate statement of the facts if it is to be credible and effective.

Someone may say that the consequences we are discussing are inevitable in a free society, that citizens of a democracy assume that truth emerges from the clash of opposing views, and that advocacy for our interest-related perceptions of the truth should be zealous and unrestrained. We concede some truth to this point of view and resist any implication that moral and scientific disagreement, including the provision of half-truths or self-serving prognostications, should be censored.

However, there are many options between tolerating lack of candor or puffery as inevitable, if unfortunate, components of public reporting on neuroscience and censorship that denies freedom of speech and inquiry. These include professional self-regulation by journalists, researchers and institutional sponsors of research. This brings us to a second kind of ethical issue that these scenarios raise.

We often think that we have duties other than those relating to the common good. These including sticking to our word or keeping our promises. Sometimes those promises are built into the social roles we occupy. For example, a member of the clergy is expected to have some loyalty to the core commitments of the tradition of which she is a member; physicians are presumed to act with their patients' best interests at heart and to be prepared to keep to themselves things told them as part of a professional encounter. Similarly, journalists claim a duty to protect the confidentiality of their sources, as well as a duty to tell the truth. A set of moral duties is often thought to be built into the structure of science. These include honesty, candor or full disclosure and a commitment to following an argument where it leads. Although sociologists are now skeptical of these claims' actual power, they remain rhetorically powerful and imply some renunciation of self-interest and commitment to the advancement of knowledge over self-promotion.

Two aspects of these role-related duties call for comment. The first is that they are based on promises or commitments that are sometimes explicit but often implicit. Young physicians may sign the Hippocratic Oath, but few young scientists read *On Being a Scientist* or some other document that enunciates a moral credo of the profession. Even fewer journalists make a formal profession of principle. But these facts do not vitiate the power of the professional duties of scientists or journalists, for they are part and parcel of the expectations of the larger society and of the professions taken as a whole. Few journalists or scientists today will pretend to renounce all ambition or self-interest. Most strive for personal success and professional advancement.

The relevant fact, however, is that within both professions there is a sense that ambition should have limits, that there are things someone might do to advance her career that are inconsistent with the role the profession is supposed to play in society. Those things are out of bounds; they are *morally* wrong – although we concede that these professionals are unlikely to use the word "morally" – because they involve the breaking of a promise between professional(s) and society.

If the professionals don't acknowledge this fact, nonprofessionals certainly do. This is why the public is so intolerant of failures of scientific integrity or journalistic fabrication: we expect members of those professions to be committed to the truth, the whole truth, and nothing but the truth.

Professional duties are based in commitments; they are also more than skin deep. We expect professional commitments to be part of someone's identity, not something she easily changes. Law faculties often describe the first year of law school as the interval in which a student "learns to think as a lawyer" – a skill that will be retained throughout a professional lifetime. The person *becomes* a lawyer in learning to think like a lawyer. Similarly learning to think as a scientist or journalist colors the way one looks at the world – with skepticism, some hermeneutic of suspicion, and a desire for hard evidence. Hopes and sentiments are to some extent bracketed.

This internalization of professional norms should mean that their violation is particularly maddening to fellow professionals. It should motivate a strong concern for self-regulation within a profession. In fact this has happened in both professions under discussion, although considerable stimulus in both cases has come from outside the profession itself.

But reading the coverage of blastomere-based stem cells and stem cell derived transplantable neurons makes one wonder just how effective self-regulation has been or may be expected to be. We think of it as the ideal form of regulation if it carries real sanctions; we do not think it can be shown to be sufficient.

In the middle of the last century Reinhold Niebuhr, theologian and political commentator, argued that, because of the inevitability of self-interest, moral ideals were unlikely to be fully realized in our common life. We could approximate them, he argued, if we committed ourselves to balancing power so as to allow social time and space for the expression of opinions that might otherwise get shut out. A mid-20th century liberal, that principle of balancing power meant for him primarily using the power of government to balance the power of large corporate interests for the sake of justice.

The issues of inadequate scientific reporting are different than those Niebuhr faced, but they have this much in common: accuracy, like justice, can easily be a casualty of the interests of researchers, institutions and reporters. Probably it is important, therefore, to assure that there is accurate and comprehensive coverage in the *public* media such as NPR, CPB, etc. It may also be important actively to work to increase the outreach of these news sources. Censorship is unthinkable; discrediting poor reporting is not. And accurate intelligible reporting will discredit hype and exaggeration.

In sum, then, we have claimed that some level of conflict among researchers, sponsoring institutions and the press is inevitable because they each have good reasons for framing their stories in characteristically different ways. These differences show up clearly in the reporting on ACT's blastomere experiments and Johns Hopkins' use of stem cell derived neuron cells transplanted into a lame rat. We think it morally important to take on these issues of exaggeration, incompleteness and inaccuracy through professional self-regulation and through provision of attractive and accessible publicly funded news sources.

SECTION

2

Secondary Gain and the Misreporting of Science: The Current Controversy Regarding Industry Funding of Clinical Research

Jerome P. Kassirer, M.D.
Distinguished Professor, Tufts University School of Medicine
Editor-in-Chief Emeritus, *The New England Journal of Medicine*
Boston, MA, USA

Jerome P. Kassirer graduated Magna Cum Laude from the University of Buffalo School of Medicine in 1957. He trained in Internal Medicine at Buffalo General Hospital (Buffalo, New York) and in Nephrology at the New England Medical Center in Boston. He joined the faculty of Tufts University School of Medicine in 1961, was named Professor of Medicine in 1974, and was the Sara Murray Jordan Professor of Medicine from 1987 to 1991. From 1971 to 1991 he was Associate Physician-in-Chief of the New England Medical Center and Vice Chairman of the Department of Medicine at Tufts University School of Medicine. Dr. Kassirer served as Editor-in-Chief of *The New England Journal of Medicine* between 1991 and 1999. He is currently Distinguished Professor and Senior Assistant to the Dean at Tufts University School of Medicine. He has also served on the adjunct faculty of Yale (2000–2005) and the Case School of Medicine (2005–2006).

Dr. Kassirer has published numerous original research and clinical studies, textbook chapters and books on nephrology (in particular, acid–base equilibrium), medical decision making, and the diagnostic process. He was a co-founder and co-editor of *Nephrology Forum* in the journal *Kidney International* and of *Clinical Problem Solving* in *Hospital Practice* until 1991.

Dr. Kassirer is a Master of the American College of Physicians and has received the College's John Phillips Award. He was named Distinguished Alumnus by the School of Medicine and Biomedical Sciences at the University at Buffalo and has received the Distinguished Faculty Award from Tufts University School of Medicine and the Distinguished Service Award of the Alumni Association of Tufts University. He holds six honorary degrees, including one from L'Universite Rene Descartes in Paris. He is an honorary member of the Deutsche Gesellschaft für Innere Medizin.

Dr. Kassirer has served on the American College of Physicians' Board of Governors and Board of Regents, chaired the National Library of Medicine's Board of Scientific Counselors, and is a past Chairman of the American Board of Internal Medicine. He is a member of the Board of the National Committee for Quality Assurance. Dr. Kassirer has been elected to the Association of American Physicians, the Institute of Medicine of the National Academy of Sciences, and the American Academy of Arts and Sciences.

In editorials in *The New England Journal of Medicine*, and in multiple publications since, he has promoted professionalism, ethical scientific conduct, patient involvement in decision making, appropriate use of firearms, and reliable approaches to the assessment of the quality of health care. His latest book, about financial conflict of interest in medicine, *On The Take: How Medicine's Complicity with Big Business Endangers Your Health*, was published by Oxford University Press in October, 2004.

CHAPTER

7

Medicine's Obsession with Disclosure of Financial Conflicts: Fixing the Wrong Problem

Jerome P. Kassirer

Our profession is under a massive delusion, namely that open disclosure of physicians', other health care providers', and clinical scientists' financial ties with industry would solve the vexing problem of conflict of interest. Virtually all the major journals, as well as a great many media outlets have jumped on the disclosure bandwagon as a solution to the complex financial arrangements between physicians and industry. In fact, the principal threat of financial conflict of interest does not stem from its surreptitious nature but from its corrupting influence on the integrity of our medical information. Unfortunately, this central, compelling theme has often been lost in the welter of brouhahas over lack of disclosure of financial ties. The requirement to disclose such ties as well as the failure to disclose them has hijacked the debate about the pernicious influence of money on the data and the expert opinions we rely on from day to day to care for the sick. We should not be particularly surprised by this diversion: it is far easier to focus on hidden, sometimes embarrassing information and cover-ups than it is on the real threat. Focusing on disclosure rather than on eliminating the conflicts is simply taking the easy way out. Disclosure is not the problem; lack of disclosure is not the problem. Bias from financial ties is the problem, and disclosure does not solve it.

1 The brouhaha over failures to disclose

Perhaps no other episode typifies the recent attention to disclosure than the revelations that *JAMA* published three articles in which financial disclosure was either absent or incomplete. These failures were not the result of lack of regulations; *JAMA*'s conflict of interest disclosure requirements were already on the books and were the subject of a two-page editorial in the journal a year earlier.[1]

It is worth recounting one of these episodes in some detail: many, however, share the same characteristics. In one of many exposes on financial conflict of interest in medicine, *The Wall Street Journal* reported that authors of a report in *JAMA* that warned pregnant women about the risks of stopping antidepressant medications were also spreading their recommendation across the country in lecture series. To some obstetricians, this recommendation was surprising because it ran counter to guidelines of their professional organizations. The article noted, "[the continued use of drugs for depression] was good news for the makers of big-selling antidepressants, who have recently faced growing questions about the safety of their medications when used during pregnancy."[2] The newspaper article further reported that the makers of antidepressants had paid most of the authors of this paper as consultants or lecturers. Yet the authors failed to disclose these financial ties to the editor of *JAMA*. The lead author explained that he didn't think it was relevant to disclose these relationships because the drug industry had not funded the particular *JAMA* study in which their report appeared.[2] *JAMA* indicated that it would issue a correction in a later issue. After *JAMA*'s editor-in-chief realized that three major failures of disclosure involved Harvard Medical School authors, she called Joseph Martin, the Dean, in frustration and said, "Joe, do something with your kids."[3] *JAMA* soon tightened its rules regarding disclosure.[4] Notably, the emphasis in this episode was on the failure of disclosure, not the sentinel problem, the conflict of interest itself.

The *JAMA* revelations followed a string of other failures to disclose. In an article in the journal *Neuropsychopharmacology* favorable to vagus nerve stimulation for depression, the authors failed to disclose their connection to the company that made the electrical stimulator. Worse still, the lead author, who also failed to disclose these financial ties, was the editor-in-chief of the journal, and he had been criticized several years earlier in the press for failure to disclose other financial arrangements. The flap ultimately led to his resignation as editor.[5–7]

Several physicians at the Cleveland Clinic had promoted a device to treat atrial fibrillation without disclosing either their personal financial ties to the device company or their institution's financial ties.[8] And the financial ties to insurance companies of two authors of a paper in the *Journal of Allergy and Clinical Immunology* that downplayed the risks of household mold exposure were not disclosed despite an explicit journal policy that required such information.[9] And 2 years after the National Institutes of Health tightened its rules on several of its researchers' ties to industry, many of its top researchers had still not complied with the disclosure regulations.[10] Most recently, the former head of the Food and Drug Administration (FDA) was charged by the Justice Department with failing to disclose his ownership of several companies' stocks, including ones that had products closely related to a panel he was chairing.[11] Over and over again, concern is raised about disclosure or the lack of disclosure.

2 The causal connection: From conflicts to bias

Once a financial conflict is disclosed, a receiver of information must assess whether the conflict has had an influence on the information. Conflicts of interest influence individuals to be biased, but having a conflict of interest is no guarantee that an individual will be biased or act on that bias. Many people forget that just because someone has a conflict of interest (financial or otherwise) is not an a priori reason for considering that person's opinion biased. Bias is often thought of as a deliberate choice of conflicted individuals. (I have a financial stake in the action; therefore I opt to bias my opinion in favor of the company who pays me.) In fact, such a causal chain is probably the least likely occurrence. Most always, the connection between a financial tie and a biased opinion is subconscious. Studies by cognitive psychologists show repeatedly that people are unaware of their biases, and that self-interest distorts judgment.[12] No matter what the causal mechanism, however, the receiver of information from a conflicted physician who declares his or her financial connections must engage in a guessing game. How do I know whether the information I am hearing is objective? How do I know whether it is biased?

Several major medical societies have weighed in on the acceptability of receiving gifts from pharmaceutical companies. Their ethics committees have generally recommended allowing members to accept small gifts that might foster patient care (such as textbooks and stethoscopes) and "modest" meals, but they are silent on consultations with industry, speaker's bureaus, and development of "educational" materials for industry. Many major medical centers including Yale, Stanford, Penn, and the medical campuses of University of California at Los Angeles and Davis have outlawed meals and small gifts, but few have eliminated these more lucrative joint ventures. The implications of any level of gift giving and acceptance are discussed later.

3 How extensive are financial conflicts?

If a small minority of physicians had financial ties to industry, we might ignore such conflicts, but such is not the case. Based on disclosure information from medical meetings, medical journals, clinical practice guideline committees, FDA advisory groups and surveys, about one-third of academic physicians have substantial conflicts,[13] and in some specialty groups the fraction may be considerably higher. At the 2005 meeting of Transcatheter

Therapeutics, for example, 44 percent of the presenters had received money from industry, many of whose products were discussed at the meeting.[14] And an ethics conference designed to set policy on conflict of interest that was sponsored by the American Heart Association and the American College of Cardiology listed 47 cardiologists as participants: nearly two-thirds had financial conflicts such as consultancies, memberships on speaker's bureaus or advisory committees, stock ownership, or other forms of equity.[15] At the 2006 meeting of the American Psychiatric Association, about one-third of the presenters had financial conflicts, but the number of senior psychiatrists with conflicts is probably underestimated by this count because many presenting were junior participants in the meeting.[16]

Numerous examples also exist of clinical practice guidelines that have come under suspicion because of heavy participation on guideline committees of conflicted physicians. These include the "andropause" guideline of the Endocrine Society, the cholesterol guidelines of the National Cholesterol Education Program, and the "clot buster" guideline of the American Heart Association.[13,17] Criticism of upcoming practice guidelines from the Heart Rhythm Society and the National Institute of Allergy and Infectious Diseases has dominated recent news.[18,19] One of the most troubling revelations deals with Xigris, an expensive drug use for patients with life-threatening infections. A practice guideline committee of the Society of Critical Care Medicine promoted the use of this expensive drug despite what other experts considered flimsy evidence of its efficacy; most of the committee members were consultants for Lilly, the company that makes Xigris. Moreover, the Society recommended that Xigris be bundled into a pay for performance program that would reward physicians for using the drug (and possibly penalize those who did not). Lilly's financial involvement in the guideline and in physicians' marketing efforts for Xigris is well documented.[20]

4 Why should I disclose? I can't be bought

Failure to disclose financial arrangements with industry may be inadvertent as when an individual simply forgets about a commercial tie. Sometimes it may be an intentional effort to cover up the financial arrangement. And sometimes physicians, fully aware of the conflict, consider it irrelevant because they believe that no amount of money can influence their opinion. Time and time again, when a speaker or author is discovered to have ignored or avoided disclosing a financial relation with industry, they offer the last explanation: they cannot be bought.

After the revelations about the hidden financial conflicts of Charles Nemeroff, the editor-in-chief of *Neuropsychopharmacology*, one of his colleagues in England offered a characteristic defense. She said, "I don't believe for a minute that the fact the paper was funded by a company would have influenced his conclusions." "It is unfortunate that he has had to stand down over this incident which is largely a reflection of the scientific community's paranoia rather than any failing of his professional integrity."[7] And when the editor of an ophthalmology journal who admitted that he had been a paid consultant for "almost every company that manufactures or sells viscosurgical equipment" complained that the *Canadian Medical Association Journal* would exclude him from authorship of certain kinds of papers, he wrote, "Like other medical editors and reviewers, I am extremely careful to avoid any possible bias in my own articles and in my reviews by other researchers. I consult for all sides on most issues; I do not care who wins an argument from the financial point of view, but I do care passionately that the academic issues are resolved honestly and correctly."[21] Perhaps the most arrogant response was cited in an article in Business Week. A New York cardiologist with extensive financial ties to industry was asked whether his connections represented a problem during an educational session in which his product was being promoted. He replied, "Absolutely not." He added that the question "borders on being offensive."[14] A psychiatrist at Columbia University who had worked on several industry-financed practice guidelines gave a pragmatic defense of such support. He claimed, "Some of the funding was simply good will by companies with no strong link to any product likely to be highly recommended." He said, "Researchers are generally making the best of a funding system controlled mostly by a commercial enterprise." "That's the market-driven system the American people have chosen to live within, and it has its pros and cons."[22]

Others are more introspective about the value of disclosure. In a published interview, Irving Weissman, a stem cell researcher at Stanford University, said, "The way I do it is I say, 'I have stock and I'm a director of this or that [company], so you might think I am biased, so you judge...'"[23] And Paul Volberding, an AIDS researcher at the University of California at San Francisco, responding to critics of a clinical practice guideline funded by a drug company and written by six physicians, all of whom had financial ties to a company whose drug is promoted by the guideline, said, "It's always appropriate to be cautious with any pharmaceutical company involvement in research or, certainly in the publication of guidelines." "It [concern about bias] can be addressed to some degree with full disclosure of the relationship to the company."[22]

The view that no amount of gifts, meals, or money can influence a physician may have its roots in our training and practices. Many of the physicians who end up consulting with industry and serving on pharmaceutical speaker's bureaus have had extensive training in scientific methods, and many who serve on committees that formulate clinical practice guidelines are heavily involved with the practice of evidence-based medicine. In both cases, fervent objectivity is the order of the day, and could easily lead professionals to believe that no matter what the financial incentive, they remain fully rational and objective.

Unfortunately, this attitude butts up against a powerful human trait, that of reciprocity. It is an important human trait that bears a close look.

5 You can be bought but you're not aware of it

Fifty thousand years ago, give or take a millennium or so, small bands of hunter-gatherers, mostly in autonomous family groups, began to appreciate that wandering around toward sources of food and away from natural and human threats was not an optimal strategy for survival. Instead they began to form small settled communities in which individuals took on specialized functions (such as water and food storage, or keeping order). Because of the new dependence on others, that is, those not in an immediate family, to carry out their assigned (or self-assigned) responsibilities, people needed to trust others. Fundamental to this new societal organization was an expectation that people acted with integrity, reciprocating in kind for the hard work of others. Such reciprocity is thought by evolutionary psychologists to be a principal cohesive force in society, and has been described as one of the "deepest wells of human nature."[24] Indeed, certain animal species, including chimpanzees and vampire bats, display striking examples of reciprocation that appears like "altruistic behavior." And in humans, altruism is thought to have one of its bases in the expectation of reciprocity for favors.

Studies by the distinguished psychologist Robert Cialdini have elaborated on this concept of reciprocation.[25] By spending time with a variety of people whose job is to sell something (e.g., salespeople, fund-raisers, and advertisers) he identified several attributes that encourage a reciprocal response. Friendly relations and social acceptance were two important factors in getting people to comply with requests from others. A reluctance to be seen as a "free-loader," or a "moocher" was another. It is surprising how small the gifts are that induce these responses; even gifts that are quickly discarded can produce an obligation. We do not enjoy being in a persistent state indebtedness.[25] Not surprisingly, a gift of food seems to induce a unique obligation for a reciprocal action, perhaps because satisfying hunger is such a primal instinct.

If evidence from paleoanthropology and evolutionary psychology is too arcane to convince some people of the power of reciprocity, recent experiments by other social scientists that shed light on self-serving behavior add to the evidence.[12,26] By devising experiments with normal volunteers who were faced with judgments concerning financial tradeoffs, investigators at Carnegie Mellon University showed that people are often unaware how self-interest influences their opinions, that they have difficulty seeing themselves as corruptible, and that when their self-interest is at stake, they find it difficult to be unbiased. Even when individuals are motivated to remain impartial, they are often unable to remain objective. This response is taken as evidence that self-serving biases are not intentional.[12]

Finally, there is ample evidence from studies in medicine that physicians are influenced by gifts and payments.[27,28] Indeed, the very fact that the pharmaceutical industry spends billions of dollars on physicians for gifts, meals, and speaking and consultation fees is evidence in itself that gifts yield substantial influence.

6 Does the amount of payment matter?

Some have argued vociferously that the requirement to disclose financial ties is nothing more than voyeurism, and an invasion of an individual's privacy.[29] In efforts to assuage researchers with financial ties to industry, some medical centers and universities have opted to allow their faculty to have such ties, but have mandated a de minimus amount, usually $10 000, that they can retain each year. The notion underlying this strategy is that sums less than $10 000 are insignificant but those greater can be problematic. Note that the $10 000 *de minimus* figure is for each company, and that individuals who consult with many companies can amass substantial sums without violating their institutions rules. Yet if we acknowledge that even small amounts of money (or in-kind gifts) engender an obligation of reciprocity, then any amount should matter. Once someone has accepted a financial payment of any kind, a conflict of interest exists. The conflict is independent of the amount of gifts, meals, or cash that generated it.[30]

7 Why focus only on financial conflicts?

All of the focus of this essay has been on financial conflicts of interest, even though I would be the first to acknowledge that in some instances other personal biases may have a far more profound effect on decision-making. Other biases are legion: religious, social, libertarian, and other ideologies and predispositions.[30,31] But there is a substantial difference between financial conflicts and others; namely financial conflicts are optional. When faced with the choice to agree to a financial relationship with a company or not to, one has a choice: either take it or leave it. In contrast, as Stark points out, "One cannot divest oneself of one's biases or prejudgments because they are so integral; one cannot easily disclose them because they are so internal."[30] In fact, as he argues, "...disclosing a psychological bias comes far closer to admitting an irremediable incapacity to make an unencumbered decision." In terms of interpreting the pronouncements of an individual, it seems nearly hopeless to ask them to disclose their non-financial conflicts of interest.

8 What's wrong with disclosure of financial conflicts?

Disclosure is based on the premise that transparency informs the receiver of information about the existence of a financial conflict in circumstances when the conflict cannot be resolved. It assumes that this information allows individuals to judge whether the conflict might have had a biasing influence on a spoken or written declaration and thus allows the receiver to discount the information. Unfortunately, it is difficult for anyone who is not a highly specialized domain expert to identify a biased opinion in spoken or written material. Even when disclosure is thorough and complete, it may not reach the appropriate audience. Such information is often left uninspected or unevaluated by authorities and not made available to those who need to know (such is often the case for FDA deliberations). Disclosure requires people to assess a physician's motives and guess whether his or her actions may have been affected by any conflicts. Some even may assume that disclosing a financial conflict is evidence at face value that a speaker is providing unbiased information. In fact, the declared conflict places the receiver of information in the position of mind reader, continually asking whether there is deliberate bias, subconscious bias, or no bias at all. I recently likened this requirement to reading Tarot cards.[32]

The details of disclosure are rarely furnished. Lecturers often flash their financial arrangements quickly on a few slides, giving listeners no opportunity even to consider the connections. Moreover, simply naming a company doesn't specify which drugs or devices the company manufactures. And because declarations of conflict of interest are voluntary, the accuracy of individual declarations is uncertain. In meetings in which all members of a group disclose their financial conflicts, accusing another group member of a conflict serious enough to exclude them is far too socially embarrassing to do. As a consequence, few do so.

I wrote, "It is a fallacy that something about disclosing, or public 'confession' of financial conflicts magically 'eliminates' bias, as though somehow it operates directly on the individual's psyche. When we announce our conflicts, we quietly and symbolically wink at each other that objectivity reigns. Instead, disclosure has come to be treated as a formality, just another piece of paper to fill out, rather than a solemn moment during which people take inventory of their integrity."[13]

9 Could disclosure actually make the problem worse?

In fact, disclosure may even exacerbate the problem because it can give "those with conflicts of interest a moral license and a strategic initiative to further skew the data."[33] Additional studies in social science with simulated estimators (of numbers of coins) and advisors show that individuals do not discount the advice they receive and make errors in judgment based on that advice.[34] The net result is that disclosing the conflict of interest makes people underestimate the severity of the conflicts. Commenting on the response of observers who had been told about an individual's conflict of interest, one of the investigators said, "You know the score, so now anything goes." "People are grasping at the straw of disclosure because it allows them to have their cake and eat it too."[35]

James Surowiecki, who writes the economics page in *The New Yorker* magazine also said, "It has become a truism on Wall Street that conflicts of interest are unavoidable. In fact, most of them only seem so, because avoiding them makes it harder to get rich. That's why full disclosure is so popular: it requires no substantive change."[35] Surowiecki's comments reflect the same phenomenon that I have referred to here, namely that disclosure is a sham; a way of deluding ourselves that we have cleansed the problem of conflict of interest and bias. It follows that the angst about lack of disclosure is much the same: a way of avoiding the really tough problem of protecting the integrity of our medical information.

Why have we so blithely accepted disclosure as the panacea for financial conflicts? Probably because cleansing the conflicts is an overwhelmingly difficult task. Institutions which accept food for their house staff find it difficult to deny food at house staff and medical student conferences or pay for the meals out of their shrinking revenues. Physicians have become so accustomed to receiving free continuing medical education, are loath to begin paying for it out of pocket, especially when the incomes of many groups are in jeopardy. Thus, if we can't eliminate the conflicts (the ideal solution), just disclose them and live with the consequences.

10 Why does it matter?

Our threshold of concern about financial conflicts must be lowered. Just because a standard is difficult to reach should not deter us from trying to reach it. Conflicts of interest lead to biased choices about which drug to prescribe and which device to implant. Biased choices raise the cost of care, may embolden researchers to move ahead too quickly and thus risk patients' welfare.[13,36] Judging from cartoons that depict doctors "on the take" and from numerous newspaper exposes about physicians who seem to have acted in their own interest over the best interests of their patients, the public could lose its trust in the commonly held notion that a doctor's professionalism successfully counters the lure of financial success. Professionalism used to be the standard by which physicians were measured, but this standard is in jeopardy, largely based on physicians' financial conflicts and the actions that flow from them. In the movie "Pirates of the Caribbean: Dead Man's Chest," a greedy pirate hunter declares, "Loyalty is no longer the currency of the realm." "What is then?" someone asks. "Currency is the currency of the realm" he replies. Professionalism has been the currency of our realm. Has it been replaced too?

11 If not disclosure, what?

I have implied throughout this essay that elimination of financial conflicts is an ideal solution. I also aver that recusal of individuals with conflicts from voting on clinical policies is not as effective as eliminating those with conflicts, and that disclosure, as currently practiced, is a sham. I have no objection to small numbers of conflicted individuals participating in discussions of policies in clinical practice guideline committees and FDA deliberations, but I would not give them a vote. Others have voiced the same opinion.[37,38]

Others disagree, and they have advanced several arguments. First, they point out that virtually all the experts are on the payroll of one or more companies and that the loss of the expertise of these physicians is problematic. Some claim that research quality and patient care will suffer from reliance on experts not associated with industry,[39] though they adduce no evidence for this argument. Some even claim that advice should not be sought from those who industry has passed by because they are not the best and the brightest.[29] The first argument, if true, means that virtually the entire academic community has been co-opted by industry: a truly depressing allegation. The last argument is elitist and condescending; industry undoubtedly selects consultants and speakers not just based on meritorious or intellectual attributes, but for traits that favor the company's marketing objectives.

I believe that medicine should adopt the same restrictive policies with respect to financial conflict of interest as other institutions that manage conflicts far better. Surely we would not hold up our politicians against this high standard (they fail it), and even many judges are known for their willingness to take trips and accept meals from companies about which they later might be required to opine. Nonetheless, one paragon does measure up to the highest standard, namely reporters. Most reporters assiduously avoid financial conflicts. If they share a meal with someone about whom they might be required to report, they either pay the tab or split the bill. From my vantage point, *The Wall Street Journal* and *The New York Times* have the most stringent policies. *The New York Times* conflict of interest policy governs the relationship between the private lives of reporters and officials and their journalistic pursuits. Among its provisions:

> No staff member may own stock or have any other financial interest in a company, enterprise or industry the coverage of which he or she

> regularly provides, edits, packages or supervises or is likely regularly to provide, edit, package or supervise. A book editor for example, may not invest in a publishing house, a health writer in a pharmaceutical company or a Pentagon reporter in a mutual fund specializing in defense stocks. For this purpose industry is defined broadly; for example, a reporter responsible for any segment of media coverage may not own media stock. "Stock" should be read to include futures, options, rights, and speculative debt, as well as "sector" mutual funds (those focused on one industry).[40]

The press assiduously avoids conflicts of interest, and when one of its own breaks with this time-honored tradition, it becomes big news.[41] Avoiding financial ties is a strong disinfectant that preserves and protects integrity. It gives a signal to others that a writer or speaker is expressing views based on unalloyed information and unencumbered judgment.

Marc Rodwin, one of the leading scholars in this field, pointed out as early as 1989 that disclosure of financial conflicts of interest is an inadequate solution to the problem, and that physicians "should not be allowed to limit their obligation to act for the good of their patients simply by disclosing conflicts of interest."[42] Disclosure can be made effective but it takes considerable effort and oversight. He later wrote, "...disclosure can help address conflicts of interest, but only if it is a part of a coordinated policy that sets high standards of ethical conduct, clearly delineates the permissible from the unacceptable, develops institutions to monitor behavior, and imposes meaningful sanctions to ensure compliance."[43] A small number of medical journals and medical societies are beginning to evolve policies that would monitor compliance and punish individuals who do not comply, but such mechanisms rarely are found in academic medical centers, community hospitals, and most professional medical organizations.

American medicine has hidden behind disclosure as a solution to the problem of mounting financial ties to industry, but the public, increasingly wary of these relationships, will demand unencumbered medical information. On subjects that could be influenced, even subconsciously, by financial ties, conflicted physicians should be excluded from participation in practice guideline committees, continuing medical education, and writing editorials and review articles for journals. Academic medical centers, medical societies, and medical journals must develop rigorous approaches to avoid financial conflicts and to deal with inevitable breaches of conduct.

Declan P. Doogan, M.B., Ch.B., FRCP (Glas.), FFPM, DSc
President, Research & Development
Amarin Corporation
London, UK

Declan P. Doogan received his medical degree from Glasgow University in 1975, and his doctorate of science degree from the University of Kent at Canterbury (UK). He is a Fellow of both the Royal College of Physicians of Glasgow and of the Faculty of Pharmaceutical Medicine in the United Kingdom. Dr. Doogan started his career in the pharmaceutical industry, as the lead clinician for fluvoxamine, for Duphar BV. He then joined Pfizer Inc. in 1982, where he enjoyed a long and highly successful career, including leading the clinical development for sertraline (Zoloft). During his tenure at Pfizer, Dr. Doogan rose to the post of Senior Vice President and Head of Worldwide Development, for Pfizer Global Research & Development.

Dr. Doogan joined Amarin Corporation (UK) in 2007, as its President of Research & Development, and he serves on the boards of several biomedical research corporations. Dr. Doogan holds Visiting Professorships of Drug Development Science with Kitasato University in Japan, with the Glasgow University Medical School, and with the School of Public Health at Harvard University.

His enduring professional interests are in identifying innovative approaches to the identification and development of new medicines and to improving the interface between the pharmaceutical industry, academia, regulators, and the public.

CHAPTER

8

In Support of Industry-Sponsored Clinical Research

Declan P. Doogan

A century ago medicines were sold without restriction. Gradually, governments required evidence of safety and for the last half century evidence of efficacy has been necessary as well. Most recently, governments and insurers have added evidence of cost effectiveness as a "fourth hurdle" requirement prior to approval of a new drug. Clinical trials are required to support all of these above needs. Such clinical research is time consuming, labor intensive, unpredictable and extremely costly. In addition to the regulatory/payer needs, there are many factors to consider, including advances in knowledge of the disease, of the compound or class of compounds, and commercial needs. The support and involvement of academics and clinicians at scholarly institutions and medical centers is also required for a large variety of scientific, political and financial reasons. The vast majority of this research is sponsored by the pharmaceutical industry and it is a vital contribution to the practice of medicine.

Society is largely unaware of the huge burden of such research and indeed its benefits. However, when things go wrong the media can be a harsh commentator on clinical trial conduct, despite the significant safeguards that are built into the clinical trial process. This chapter will discuss these factors in detail as well as describe how new compounds are developed.

1 Disease knowledge

The industry must judge the merits of its products against a background of understanding how the disease is classified, how it evolves over time and its response to existing therapies, including placebo. In some diseases, such as depression, the placebo response rate can be profound, further adding to the complications of showing efficacy. In this particular condition it is not uncommon for double-blind studies of new agents to fail to show a difference from placebo. This does not necessarily mean the compound is ineffective; failure to show efficacy can be due to many other non-drug factors. Because of this variability of treatment effect over the past 20–30 years double-blind placebo controlled trials have been extensively used to verify the treatment effect. Without placebo control, trials would have to either enroll many more subjects or active treatment might be inappropriately considered effective (the type II statistical error).

Trials are now usually designed with sufficient power and precision to answer the research question based on advice from biostatisticians. This information can then be integrated with knowledge of the prevalence of unwanted symptomatology which could be misinterpreted as drug-induced side effects. During the course of such research new knowledge of disease patterns, characterization of subgroups and unknown effects of existing treatments may also emerge. Given the number of studies conducted, meta-analyses can be employed to better describe disease responses to particular therapies, including gaining a better understanding of already approved treatments. The recent withdrawal of **Vioxx** led to a large number of meta-analyses conducted not only by sponsors but also academics and regulators. These contributed significant knowledge of the background incidence of cardiovascular morbidity in rheumatic conditions and the safety of existing non-steroidal anti-inflammatory drugs (NSAIDs). These agents had been on the market for many years and had been prescribed to many millions of patients. Indeed, there had been no systematic review of the cardiovascular safety of the NSAIDs until the withdrawal of **Vioxx**, as it was presumed there were no such safety issues given their time on the market.

2 How is clinical research conducted?

Once a compound is perceived to posses the necessary preclinical characteristics, including pharmaceutical and animal safety qualities, and thus is ready for human evaluation, the first stage of clinical research is called Phase I. Usually the first study is in a small group (20–100) of healthy volunteers to determine the compound's safety profile, the safe dose range and pharmacokinetic properties. There are usually preset thresholds which must be passed before the compound is permitted to progress into patient studies. Studies conducted with patients are included under Phase II. These studies usually involve double-blind studies with placebo control (except in life threatening conditions such as oncology). The goal of these studies is to establish the "proof of concept," where the medicine is shown to beneficially impact the disease and has a suitable benefit–risk ratio. If the product continues to posses the required characteristics for additional clinical research, the next stage is called Phase III. Now large, randomized, double-blind controlled trials with much larger numbers of subjects are initiated in order to gain a much better understanding of how the drug behaves in a broader population. This population usually consists of outpatients; however this is dependent on how the disease is managed in usual practice. During the entire program the composite body of knowledge of the product is analyzed and reported to show the benefit–risk ratio and provide regulators, prescribers and the sponsor confidence of its medical utility.

At all stages there are procedures to ensure the correct conduct of the clinical experimentation. The sponsor must be satisfied the appropriate information has been gathered to permit the progress of the compound. Regulators receive data packages under the Investigational New Drug (IND) procedure in the United States. This begins with the submission of preclinical data, which are reviewed together with the first human study protocols. The Food and Drug Administration (FDA) must give approval before human testing can begin. In addition to this, each trial must have a suitably qualified principal investigator who is responsible for the ethical and medical conduct of the research and is ultimately responsible for the safety of the trial subjects. Finally, the Institutional Review Board (IRB), which comprises individuals charged with ensuring that the study is appropriately designed to answer the trial objectives, must make certain that sufficient safety safeguards are built in. These individuals have no vested interest in the study and are capable of providing independent guidance to the institution where the trial is to be conducted.

3 Regulatory drivers

The development program is established with the regulatory authorities in advance. The new drug application (NDA) is the primary filing necessary in order to enable the FDA to review the product characteristics. The NDA is filed when the agreed program has been completed by the sponsor who must be satisfied that the product characteristics are met and the product stands a chance of gaining acceptance by prescribers and patients. The product will be approved on the basis of fulfilling certain levels of safety, efficacy and product quality. This process takes on average about 17 months and between 10 and 15 percent of applications are rejected (see: www.phrma.org). There are US and European guidelines for many indications but each product must have its program individually designed due to its unique characteristics and many highly innovative candidates are filed for indications for which there are few or no regulatory guidelines. During the review process there is often negotiation between the sponsor and the reviewer due to differences of opinion as to how the data are to be interpreted. This can lead to regulatory delays, non-approval letters and/or the need to conduct more studies before approval. When successful, a product label is agreed upon, which informs prescribers by describing the characteristics and use of the product, as well as alerts them to potential safety issues.

Trials are conducted on approved and marketed products for a number of reasons. Product label enhancement is a necessary means of adding new knowledge of the product in terms of safety, new indications and formulations. For example, atorvastatin (**Lipitor**) was originally approved for familial hypercholesterolemia and is now a broadly used product for the control of hypercholesterolemia as a result of large and expensive trials carried out after approval of the initial indication. Some trials are regulatory obligations, for example those for safety in patient subgroups while others, for example new formulations or new indications, are usually driven by the sponsor's wish to increase market share. *Pharmacovigilance*, the term used for the monitoring of safety signals, can include a review of spontaneous adverse reports and at some point may lead to a label amendment.

4 Commercial drivers

New indications for existing drugs can be identified from new biology or pharmacology discoveries, serendipitous clinical findings or literature reviews. In any case these new findings are usually underpinned by further clinical research and, if of sufficient medical importance, regulatory review.

Another driver for clinical trials is marketing support. There is a tension in the industry between rapid recruitment of patients to clinical trials, the quality[a] of those patients and the need to secure key opinion leader (KOL) support. Experience teaches that many KOLs often are not responsible for assessing many of the patients recruited from their centers owing to their workload demands. Also recruitment rates can be compromised as there is competition for the same patients from several research compounds, investigators and study sponsors in the clinic at the same time. If the sponsor does not participate in this system that is managed by the key medical experts, then these influential figures will be less likely to endorse the product and hence access to a broader patient population will be more difficult. (*The challenge is to identify KOLs who are both thought leaders in the particular disease and have the research experience in the use of the product to enable their support.*) To compensate for recruitment difficulties, companies cast their trial nets wider including other countries/regions which are considered non-traditional, such as Eastern Europe, Latin America and the emergent Asian countries.

[a] I use the term "quality" to mean the appropriateness of the patient for inclusion in the study. Rapid recruitment can often increase the number of borderline or unsuitable patients – those that might not fully meet all inclusion and/or exclusion criteria for subject selection in a study.

From the investigator's perspective, participation in clinical trials is a good way of ensuring a revenue stream for their institution. Furthermore, presenting the results at a variety of scientific gatherings, and writing peer-reviewed scientific publications based on these clinical trials, are both core activities of KOLs.

5 Regional needs

This leads to consideration of regional differences in medical practice. Despite the agreements struck by the International Conference on Harmonization (ICH) and Good Clinical Practice (GCP) regulations there is still the potential for extrinsic factors such as diet, medical practice and the reimbursement system to influence therapeutic outcome. At a regulatory level it is highly undesirable for all the data to be derived from a single region, even the United States. This can lead to repeat of studies of a similar design across several regions. In one way this can lead to redundancy but it also leads to the accrual of a larger body of knowledge of the products. A particular issue for product approval in Japan is the need to repeat a significant amount of clinical research in Japanese. This requirement is based on concerns that Western data are not representative for Asians as they are genetically different. These so-called intrinsic differences are thought to lead to differences in pharmacokinetics and in some cases drug effects between the West and Japan. In addition, extrinsic differences (e.g., medical practice, disease classification and diet) are believed to add to the difficulties of using one region's data as representative for the other. With experience and harmonization of trial methodology it would appear that extrinsic differences do not significantly influence results. There are some real intrinsic differences that must be accounted for but these are dependent on the molecule and should not be considered to be systematically present. In studies of interethnic drug metabolizing differences between Asians and Caucasians, it has been shown lifestyle factors exert negligible effects. There are some intrinsic differences for such associations between genotype and the drug metabolizing capacity of the cytochrome P450 system CYP2C9, 2C19, but not for 1A2, 2E1 and 3A4/5. Implications for these data include: (a) possible interchangeable use of Japanese, Koreans and Chinese subjects for drug development in these countries and (b) possible use of ex-patriated Japanese subjects (e.g. USA-based) in registration trials for submission to the Japanese regulatory authorities. This provides practical advice to regulators and sponsors in study design and interpretation.

This leads to an ambiguity in the interpretation of data. Regulators are comfortable reviewing safety information combined into a single body of experience. This is especially necessary for infrequent adverse events. The ability to find rare severe side effects is solely a function of treating large numbers of patients for long periods of time. For benefit, on the other hand, there is a much greater reluctance to accept the results as poolable because of the regional/genetic differences as mentioned above. This is not totally defensible and there is a certain degree of implausibility to this prejudice. This again results in the repeat of studies of a similar design which are usually confirmatory. The fundamental difference between the two is that efficacy parameters are defined a priori while safety signals are often just observed as a result of the numbers of patients treated.

Most diseases recognized in the developed world are diagnosed and classified in a similar manner across regions thanks to the significant efforts of the medical and regulatory communities. This enables the development of similar formularies around the world. There are, of course, exceptions which may be political in nature, for example the oral contraceptive was only approved in 1998 in Japan. It coincided with the rapid development and approval of sildenafil (Viagra) for male erectile dysfunction. The delay with the oral contraceptive was supposedly due to safety concerns. This is remarkable on two counts in that the same preparations had been available in the West for 40 years and sildenafil had been available internationally for less than 3 years. The politics of product acceptance also drive clinical research strategies.

6 Pricing and reimbursement

Because of different health care systems, reimbursement decisions are not uniformly applied. In the United States the so-called fourth hurdle is not applied pre-NDA approval. Negotiations take place with payer institutions for discounts to enable formulary acceptance and reimbursement. Each payer can negotiate whether to accept the product on the formulary and may need different evidence to convince them of this. In the United Kingdom, the National Institute for Clinical Excellence (NICE) is charged with determining the evidence which can support National Health Service (NHS) reimbursement, though not determining the price. This is in addition to the data required from the Medicines and Healthcare Products Regulatory Agency (MHRA) for approval of the product label. This leads to an added burden of proof and ultimately gathering of more data. In most countries there is a serious lack of epidemiological data that characterizes diseases with sufficient specificity which can show the baseline effects of existing treatments that enables the new intervention to be measured against. There is an absence of systematic guidance and global alignment on standards for not only conduct but need for pharmacoeconomics and outcomes research studies. Neither is there agreement on the degree of efficacy the new treatment must have to be considered not only clinically, but economically significant. Regulators and payers do not wish to agree on a pre-set threshold of product performance that would warrant approval or reimbursement. This would compel the payer to reimburse the product or the regulator to approve if those criteria were met. This uncertainty drives imperfect study design and execution. It is the equivalent of shooting at moving target.

7 Attrition

It is estimated that only 1:10 000 new chemical entities become medicines.[1] Most attrition takes place preclinically but once these agents enter clinical development the survival is still extremely poor. The survival rates in Phases I, II and III/Approval are 60, 25 and 70 percent respectively. Compounded, this leads to an overall survival in the clinic of around 10 percent. This is not to say that the industry engages in wasteful research but it is clearly a sub-optimal system that allows such efforts to be expended on compounds that will never become medicines. There is much to be learnt from these experiments in order to refine the study designs for future compounds. With such a poor survival rate it is not surprising that once an agent has demonstrated a sufficient benefit–risk to warrant regulatory approval there is a desire to capitalize on all therapeutic possibilities for the product. Here it depends on the vantage point of the commentator as to the need for subsequent studies. For example, a new formulation may not have the same impact on the payer as it does the patient. A new product that delivers the same efficacy as an older, less expensive therapy but is either more convenient or has fewer non-serious side effects may be very attractive to the patient. How does the payer value that benefit? Conversely, when a product is used off label for an indication and the sponsor does not conduct a program to systematically study this, it will be criticized and reimbursement will be difficult.

8 Adding to the body of knowledge

Contrary to popular belief, the majority of clinical research is sponsored by the pharmaceutical industry. Although the National Institutes of Health (NIH) play an important role in biomedical research in the United States, spending $28.6 billion in 2005, the biopharmaceutical industry spent $51.3 billion in the same period.[2] Globally, this figure rises to $60.4 billion.[3] In addition, the NIH-sponsored clinical trial spending of $2.78 billion in 2004 compared to $4.99 billion from industry.[2] This is particularly important to note in the conduct of mortality and morbidity studies. These enormously expensive undertakings have led to a significant increase in our understanding of key aspects of the disease especially in review of cardiovascular outcomes data. The investment in the statins over the past two decades has enabled these compounds to be used broadly with a significant impact on the morbidity and mortality of cardiovascular disease. This must surely be considered a triumph of modern medicine and a wonderful example of a coalition between academia, clinical medicine and industry.

Recently the torcetrapib/atorvastatin combination developed by Pfizer was terminated because of an increased death rate in the experimental arm compared with atorvastatin alone. This was only discovered after the full clinical program had been set up and 16 000 patients had been enrolled in a morbidity and mortality study. It was unlikely that this program could have provided the same answer without the large patient exposure. Had the product shown the requisite benefit–risk it would have been an example of front-end investment and enabled extensive usage early in the post-approval period. By contrast, many drugs are "built in the marketplace," that is, by an extensive trial program conducted after marketing authorization and when a revenue stream can help to fund the research program. These investment strategies must be consistent with the financial ability of the company to pay. It is quite understandable that companies would prefer to delay, as much as possible, trial investments until after product approval. But many, like torcetrapib could not have been approved without a massive pre-approval investment. This is but one example of the huge investment risks that the industry takes all the time.

9 Is the clinical research budget sustainable?

Research and development costs are significant, and they are most definitely rising each year. A recent survey by the Tufts University Center for the Study of Drug Development (CSDD), published in 2006, reported that the average cost to bring a single compound to the point of submission of a "NDA" to the FDA was $1.3 billion in 2003, reflecting an increase of nearly $500 million compared to 2001 figures.[4] Biologics are similarly expensive with an estimated cost of $1.2 billion per Biologics License Application (BLA). It has been estimated that the annual R&D expenditure by the industry increased by 147 percent from $16 billion to $40 billion since 1993.[5] If this could be realized less expensively, then the industry would certainly do so, but the drivers for such large clinical research budgets exist both pre- and post-marketing authorization. Assuming a product is successful, the size of the program is dependent on the nature of the disease and the treatment modality. For example, a single oral dose treatment with little investigation or monitoring will be significantly cheaper than a chronic use agent with costly and invasive monitoring. This will likely impact the down-stream revenues of the compound.

One of the greatest challenges that the industry faces is to better determine, as early as possible in development, which compounds are likely to not progress past Phase II of drug development. Figure 8.1, clearly shows the origins of the decline in Phases I and II where the negative trend has

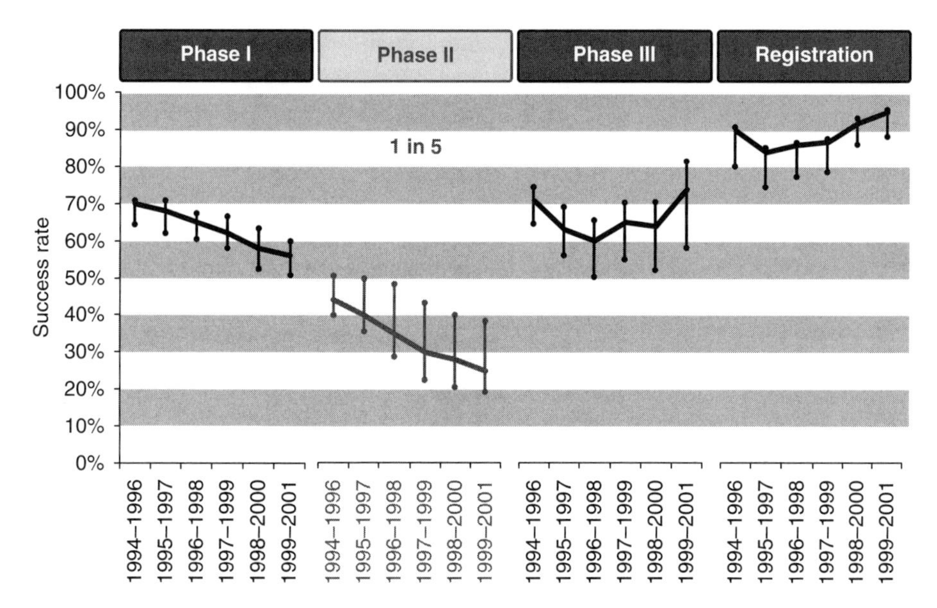

FIGURE 8.1 Candidate survival by phase of clinical drug development. Data averaged across pharmaceutical industry, and available at www.phrma.org. Data adapted for display by Dr. Doogan.

been obvious for the past decade without any signs of amelioration. Once the product has passed the early hurdles it seems as though output is relatively constant. However, the failure of a product once approved is devastating. This affects not only the patient population but also the company, which loses not only the sunk R&D costs but the significant launch expenditures.

Post-approval commitments are agreed on with regulatory bodies as a means of securing earlier marketing authorization. Usually these commitments are necessary to address residual safety concerns. These can be quite onerous and extend over many years when they address long-term safety issues. All of this leads to burgeoning development programs and companies naturally attempt to maximize the return on this research investment by in some cases making a virtue of the research effort publicizing the activity and also producing interim analyses.

10 Who are the beneficiaries of sponsored research?

The public benefits first and foremost by the availability of new medicines. It is often taken for granted that the pharmaceutical industry contributes these new medicines having discovered and developed them at great risk, difficulty and expense. The industry is also a major supporter of academic institutions around the world by providing financial support. This financial support may also sustain non-drug studies such as research fellowships, educational grants or equipment provision. The reason companies do this is to ensure a positive relationship with the institutions and the opinion leaders or prescribers. Unfortunately, this can be perceived negatively by hostile commentators. In reality, if such sponsorship were not available, many research projects would cease and many junior researchers would not be funded.

Institutions can defray some expenses by charging for the use of facilities and salaries of research staff, who also can provide standard healthcare in addition to their research commitments. The medical fraternity can access new medicines and the associated technology advances these medicines bring. Patients benefit from earlier access to these medicines provided free of charge during the trial program. Their treatment costs are also subsidized by the sponsors.

Also, new knowledge is created which can advance medical care, new techniques can be developed and promulgated and a whole industry is built around the publication of research findings.

11 Fraud in clinical research

We must draw a distinction between willful falsification of data and judicious inclusion in publications that could alter the interpretation of the findings. Journals are major vehicles for the dissemination of trial results and need to be confident in the integrity of the data being reported. Recently, the editors of *The New England Journal of Medicine* and *The Journal of the American Medical Association* publicly decried Pfizer and Merck for their conduct of their COX2 programs (Harvard-Schering Plough Symposium, June 2006). This was not about falsification of data, but rather the issue at hand focused on these companies being selective in data reporting (a claim both companies vigorously deny). Over the years the industry has been criticized in the manner in which it conducted trials and also published the information. If the data do not support the hypotheses or do not show the product in the best light there is a sense of the sponsor's reluctance to be fully transparent. One defense is that negative or inconclusive studies are not considered attractive publications to journal editors. This can be managed by electronic publications or sponsor web sites and new disclosure and transparency guidelines espoused by the industry and increasingly by legislation.

As stated previously, the integrity of the data can be assured by the research being conducted within the provisions of "GCP."[b] This permits a

[b]GCP refers to a set of procedures that have been agreed as a common standard for clinical research across the industry. They ensure studies are conducted to preserve the interests of patients such as safety, confidentiality and the ethics of study together with the integrity of the data.

record of not only the clinical data but the processes undertaken to collect it. This is the source medical data and the data trail containing the exact checks of the data during the monitoring process. For example, if the procedure requires that the patient be seen by the physician every week during the study and on monitoring this was found not to be the case, a deviation report is created. This event is reviewed before the data are considered to be not only final but meet the protocol requirements. Should the deviation be significant enough to compromise the data, then it may be excluded from the analyses. All of this is recorded to permit the regulator, during the review, to be assured of the integrity of the study. The data are subject to frequent audits by the companies themselves and by regulators. If the data do not stand up to scrutiny by audit then they will not be included as primary evidence of product benefit–risk.

Unfortunately, academic led research does not receive the same degree of scrutiny. There is no mechanism to undertake the audit of academic led studies which may also be fraught with not only procedural mistakes but data inaccuracy. Of course, there have been clear cut cases of fraud and academic research is not without its contribution. The most spectacular recent example of abject fraud was the South Korean report of successful human cloning.

In the case of the industry, there is a symbiotic (sometimes love–hate relationship!) between sponsors and journal editors. The perceived lack of transparency conveys an impression of having something to hide. There is a suspicion that companies deliberately hide data that do not serve their needs. This is generally untrue, though in the past there have been examples where studies were not reported to the satisfaction of journals, academics and the media. In the face of this, the industry is now, albeit reluctantly, publishing results earlier and also detailing where trials are ongoing on the web site www.clinicaltrials.gov. This did not happen spontaneously. It took a great deal of public criticism before industry agreed to this greater degree of transparency. This was not secrecy for secrecy's sake; there were concerns over loss of commercial advantage and increasing liability to litigation. This has not yet proven to be the case. Curiously, academic-sponsored research does not come in for the same criticism. Nevertheless, there is genuine suspicion in the minds of many stakeholders. Whilst not fraudulent in itself, the apparent judicious selection of data can introduce a bias in interpretation of results. If GCP was the remedy to assure the integrity of the data an equivalent of GCP for publications would go a long way to solve the integrity of publications.

Another type of bias is the tendency to publish only positive findings. Inconclusive or negative results are considered the least valuable and hence the literature is unlikely to be totally representative of the accumulated body of knowledge of therapeutics or disease. This bias affects not only industry but also academic research. There have been a few recent efforts to remedy this sort of bias in scientific reporting. For example, the *Journal of Negative Results in BioMedicine*, a web-based publication, was created in 2002 to publish papers on "all aspects of unexpected, controversial, provocative and/or negative results providing scientists and physicians with responsible and balanced information to support informed experimental and clinical decisions." The Editor-in-Chief is Prof. Bjorn R. Olsen from Harvard Medical School, and I view this journal as a very welcome step in the right direction.

12 Conflict of interest

Industry sponsorship is the lifeblood of commercial and academic symposia and congresses. These symposia are costly events and are a gathering place for professionals with an interest in the therapeutic area. Not only do these professionals get the latest research information, they have the chance to meet and exchange ideas for future research. As an example, cancer research has grown significantly in the past decade and many new therapies are becoming available due to the explosion in new knowledge of molecular biology. Because of the intense interest in such developments, the media are keen to learn the latest developments. At the 2006 American Society College of Oncology (ASCO) meeting, with up to 25 000 delegates attending, there were daily bulletins on CNBC's *Squawk Box* as the emergent news could significantly impact the stock valuation of pharmaceutical companies. In addition, because of the public interest other media report these events regularly. This leads to stringent safeguards being instituted to prevent advanced "leakage" of information that might have a material impact on a company's share price.

Clinical research takes place in an ecosystem. This complex world has matured over the decades into a sophisticated enterprise with much at stake for the key stakeholders. Virtually everyone is impacted by this work and it is surprising how little understood it is. Compare that with the activities in the financial sector where there has been much media attention. Enron is a good example that shows that humans can succumb to manipulating the system for personal gain. The clinical research world has its checks and balances as does the financial world. The outcry at the frankly illegal activities led to the Sarbanes-Oxley Act. This significantly added to the bureaucracy needed to assure fidelity of financial dealings as did GCP in clinical research. The research world does not have the equivalent of the Securities and Exchange Commission (SEC) to ensure the integrity of the usage of information. However Eliot Spitzer, the New York Attorney General, has successfully prosecuted several companies for not taking appropriate action to inform the health care professionals and the public about emergent data and how it may impact prescribing decisions.

13 What would happen in a world without industry-sponsored clinical research?

There would be a total dependence on academic institutes to conduct the trials and to have the requisite infrastructure. This would require funding from alternative sources. This would have to come from government or payers. It is estimated that in the United States the annual clinical trial budget of industry is approximately $5 billion. Little of this would be saved, though it could be argued that unnecessary clinical trials aimed at securing KOL endorsement, but not really adding meaningful new knowledge, would significantly decrease.

Such a model would necessarily require new systems to administer the need for data, the priority and the resource to undertake the studies. In fact it would take all the operational elements of the clinical trial process already in place in the industry and replicate them. The challenge would be to agree on these priorities as the free market would not operate. Also, the payers would be unlikely to conduct these studies as there is no incentive in their business model. Would there be a global system to share the burden? This would be extremely burdensome and again require the creation of a new bureaucracy to administer.

14 Use of new technology

The acquisition of new knowledge using traditional clinical trial methods is extremely costly, resource intensive and inefficient. In the "Google" information age, knowledge can be acquired using web technology and the connectivity between stakeholders is changing. For example, when subjects participate in a trial and their responses are measured, a study report usually emerges which finds its way to a regulator as part of a dossier or as a publication. This labor intensive mechanism is in need of significant reengineering. Data gathering is still predominantly done by pen and paper; there are multiple re-entries of analog data. Much of this could be improved by electronic health records (EHR). Observational studies, especially post-approval, could be conducted much more cheaply if EHR were available. This would permit HMOs and national governments to undertake such research independent of the industry and perhaps change the manner in which decisions on products are made, such as in terms of medical outcomes and identification of patient subgroups.

15 Conclusions

It would appear that, despite limitations in the present system, there is no viable alternative. These limitations acknowledged are being addressed with

(a) greater transparency with the publication of ongoing trials on the ct.gov web site and more rapid release of results (b) new technologies such as EHR and (c) the self-monitoring of industrial practices by PhRMA.

The clinical research process is complex and becoming increasingly sophisticated. As it is an activity that has many participants, it requires a huge effort to assure its integrity. Because of its impact on the health and welfare of individuals there is intense interest in how it is conducted and the results of these research projects. There is a clamor for new medicines and the only practical source for these is the pharmaceutical industry. There must be a critical partnership between the industry, the health care system to study the diseases and the regulator to approve the products with the appropriate benefit–risk characteristics. This partnership should not be one which the public would be suspicious of (i.e., a cozy relationship between the key stakeholders). It must withstand scrutiny and have the necessary degree of transparency to allow optimal treatment decisions. Other contributors such as scientific journals must also recognize their responsibility to publish fair and balanced information and recognize when they have a vested interest of their own. The industry must put strenuous efforts into regaining the trust it has lost with the public believing it has somewhat cynical motives to withhold product information that is not supportive of its products. Academia must also recognize the benefits of new product development and it should ensure the integrity of its own practices. There is a significant tendency for academics, regulators and payers to expect the industry to absorb the increasing costs of administering the trials (presumably because of the huge profits they make). However if the costs of clinical research are not contained, new products will not be developed because the financial burden is too great.

That drug development is complex, the rate of failure is unacceptably high and the uncertainty surrounding regulatory approval and reimbursement make this a hugely risky business. In this trial ecosystem everyone can have an impact and all decisions have consequences. If the aim of the enterprise is the discovery and development of new medicines we all stand to gain. Our future health depends on it. We should ensure all stakeholders are well informed and care about ensuring the success of clinical research. A society that explicitly values such work will ensure it has an influence on how it is conducted.

Acknowledgments

I am indebted to John Arrowsmith and Steve Williams from Pfizer Global Research and Development for their support in the preparation of this chapter.

SECTION

3

Social Pressures that Impact on How Science is Reported

The Very Reverend Dr. Wesley Carr, M.A., Ph.D., K.C.V.O.
Dean Emeritus, Westminster Abbey
Church of England
Hants, UK

Dr. Carr was Dean of Westminster from 1997 until ill health forced his retirement in 2006. During that time he was responsible for Westminster Abbey, where among many major services were the funeral of Princess Diana, the Queen Mother's funeral, and the memorial service for those who died in New York on September 11, 2001.

A graduate in classics, ancient history, philosophy, and theology from Oxford and Cambridge (and with a semester in Geneva), he was ordained in the Church of England in 1967. After a curacy in Luton Parish Church, he returned briefly to Ridley Hall, Cambridge, to teach New Testament. Subsequently he was elected Sir Henry Stephenson Fellow in the University of Sheffield, where he was awarded his Ph.D. He moved to Chelmsford Cathedral as Deputy Director of the Cathedral Centre for Research and Training, Director of Training for the Diocese of Chelmsford and a Canon Residentiary of Chelmsford Cathedral. In 1987 he was appointed Dean of Bristol. His main ministry has been consultancy, education, and training throughout the Church of England, and with other institutions and organizations both in the United Kingdom and elsewhere. Other appointments include an Honorary Fellowship at New College, Edinburgh, where he was instrumental in setting up the Media and Theological Education Project. In 2006 he served as the Erikson Scholar in residence at Austen Riggs, Stockbridge, Massachusetts, and he is a member of the Council of Scholars of the Erikson Institute. He holds honorary doctorates from the University of the West of England (Bristol) and the University of Sheffield.

Since 1974 Wesley Carr has been closely associated with the Tavistock Institute of Human Relations Group Relations Programme, and he has directed a large number of group relations conferences in Great Britain and the United States, including the prestigious Tavistock "Leicester" Conference.

Dr. Carr is series Editor of the New Library of Pastoral Care (SPCK). He has published a number of books, including six on pastoral studies. With Edward R. Shapiro he jointly wrote *Lost in Familiar Places* (Yale University Press, 1991), and he edited a collection of essays offered to the new Archbishop of Canterbury, *Say One for Me* (SPCK, 1992). In 1992 he wrote the Archbishop of Canterbury's Lent book, *Tested by the Cross* (Fount, 1992). He was general editor of *The New Dictionary of Pastoral Studies* (SPCK and Eerdmans, 2002). He has also written and published many articles and reviews. Most of these address his focal interest, which is the relationship between the human sciences, theology, and the Church's ministry.

On his retirement Her Majesty The Queen made him a Knight Commander of the Royal Victorian Order, and the Dean and Chapter of Westminster appointed him Dean Emeritus.

CHAPTER

9

Prevailing Truth: The Interface Between Religion and Science

Wesley Carr

Magna est veritas et praevalebit: "Truth is mighty and will prevail";
Or as with some justification it has been translated:
"Truth is mighty and will prevail a bit".

1 Précis

It is still widely reckoned that two large and generalized entities – religion and science – have always been, remain and are condemned to be at odds. Science is believed to have undermined religion and religion itself is regarded as unscientific. It is my belief that the falsity of each position is evident.[1] I suggest that the relation between the two is better thought of as an oscillation, moving closer and then further away, depending upon the cultural climate, not something intrinsic to them. The work of Prof. José M.R. Delgado in the 1960s, and described earlier in this volume, illustrates issues of integrity, authority and truth, which are among the marks of both faith and scientific research. All of us have an interest in preserving both of them, but who takes responsibility? One potential way of encouraging a change in this attitude is proposed. A new factor in this living relationship between science and religion is the modern media. The essay concludes with some attention to the effect of this new phenomenon on subjects with distinguished traditions such as science and religion.

This book appears at a nodal point in history when confidence in the sciences is waning and is not being found anew in formal religion. This essay is a more general contribution than the others and invites reflection on aspects of the continuing relationship between science and religion and its impact on life today. The third and new factor in this discussion is the media-saturated context in which this relationship is worked out.[2]

2 Oscillation and divergence

Both "science" and "religion" are large concepts. So issues raised by this study, and not least by Delgado himself, need to be considered in their social and historic context. The events did not just happen: they occurred at specific moments and need to be evaluated in relation to these. Science and religion are also twinned, sometimes indissolubly, sometimes far apart. Their respective histories are of oscillation from closeness to distance. Here are two examples.

In the ancient Greco-Roman world there was little difference, if any, between techne (skill) as "inspiring to design" and as "skilled to make," whether the outcome was to be a memorable statue or a simple piece of furniture. The dramas of the theater and religious activity, which were synonymous, were also skills. Most Greeks being familiar with the sea, knew instinctively that everything flows[3] and that movement was the essence of life. Indeed, the art of living was itself a kind of techne. Science was not separate from everyday life and religious practice was far more public and social than today.[4] The two lived for generations within the same frame of reference.

Much later, at the end of the 16th century, an Italian produced one of the most significant inventions in the history of the world, the telescope. This instrument reversed our way of seeing things. Galileo (1564–1642) opened up the external world and gave people a new perspective on their environment. This, together with his major offence (belief that the earth revolves round the sun), was a controversial extension of human awareness of the transcendent. It not unsurprisingly frightened the church. Only as recently as 1993 has the Vatican "acquitted" Galileo and allowed people (theoretically for the first time) access to his thought.[5]

3 Toward modern science

From about 1660 a major oscillation steadily moved science and religion apart. The prevailing Deist theology was modeled on Paley's watch.[6] The Almighty had set the world running, as it were, like a clock; and he presented himself through the design of the creation. It was an era of exciting discovery in science and complacency in religion. In Europe the churches had gone through their upheaval a century earlier in the reformations. But in fact the two now drifted closer together. So the 19th century began in a fairly somnolent state: scientists were discovering things and beginning to pursue an experiential approach to chemistry (and naturally alchemy), mathematics and physics, as well as cataloging and investigating the attributes of flora and fauna and nature in general. Meanwhile the church was occupying its place in society. Religion and science were on the whole contentedly together. What was to come could scarcely have been imagined.

Between 1840 and 1960 change was radical. A fundamentally different relationship between science and religion emerged which became an open struggle. Theology, Queen of the Sciences[7] (although her golden period had been that of mediaeval Christendom), finally surrendered and the scientific method of experiment and discovering and codifying laws reigned supreme. Individual scientists and churchmen remained friends, but structurally public debate and science in the mid-19th century (and in particular the question of evolution) paralleled in fervor the beginnings of the 21st century and the debate over stem cell research.

Similar feelings are found in each group as the church's path and that of science diverged, with the first use of the term "scientist," a label that afforded professional identity to a role that hitherto had been largely the privilege of the privileged, dating from 1834.[8] The scientific agenda was irresistible. People were overwhelmed by the new studies (as yet they could not be called "sciences"). Robert Browning put it thus:

> How you'd exult, if I could put you back
> Six hundred years, blot out cosmogony,
> Geology, ethnology, what not,
> (Greek endings, each a little passing bell
> That signifies some faith's about to die),
> and set you square with Genesis again.[9]

From time to time religious authorities could tweak religious practice. But each time it seemed yet another surrender to the cold, inhuman analysis of the new high priests – the scientists. All, whatever their stance or belief, had to surrender themselves to greater uncertainty and ambiguity: Are we basically monkeys or men? How does the unconscious life impinge upon our conscious behavior? Am I only a mouse trapped in a machine called the state? Everyone was caught up in the struggle, from the highest to the lowest and understandably from that period our current narcissism derives.[10]

It is instructive to note that what today we call "behavioral sciences" were not easily assigned a place among the sciences. The questions were whether psychoanalysis, psychology, sociology and anthropology and the like were subject to the same rigor as other sciences. If they were not, then their place among the sciences was questionable. Much of the 20th century was spent trying to find the answer, resulting in the massive intellectual and emotional investment that these disciplines represent. While they have not wholly lost their therapeutic dimension, they have taken on a significant role among the disciplines that are collaborating to develop contemporary understandings of the brain. They indeed may better be considered "human sciences" than "behavioral." And here we touch on the century-old debate that ultimately concerns our core identity.

Three intellectual giants of the 19th century between them cover most of the ground, and each addresses the individual. They are Freud, Darwin and Marx. Firstly, there is "you" with your individual psyche. Sigmund Freud did not discover the unconscious world of the self. Dreams, for him "the royal road to the unconscious," had been interpreted by mystics and priests. But through psychoanalysis Freud showed that our inner world could be ordered and interpreted as an approach to healing. Secondly, there is "you" within and between species – Charles Darwin's contribution for some remains controversial, although few now doubt its basic thrust.[11] Thirdly, Karl Marx offered a vision of a society in which each individual knew security. Paradoxically, however, it has been the experience of those in a number of countries, where during the 20th century it has been tried, that it has failed to produce such a hope for society. Each of these three represents a major discipline that has grown more complicated, so that the problem of holding together the various aspects of "you" becomes more acute.

These three figures have much in common. Each has been both demonized and beatified. Their influence on the world has been greater than the specific scholarship that they developed. And, as it has grown, so the need for these three to be right in matters of detail seems to matter less, except perhaps to small coteries of fundamentalist scientists. In passing, recent major developments in physics and cosmology seem not to have a champion, although perhaps Stephen Hawking might qualify.

The impact of such science may itself be regarded as religious, since it will affect people's awareness or sense of their place in the universe and some working on whether it is meaningful or not. The dominance that the various sciences exercised early in the 20th century has not quite gone. But no longer has it its original force. Taking broadly the main three broad areas of science – chemistry, physics and biology – we may note that in the United Kingdom recently two universities have closed their chemistry departments because of lack of student interest. Physics becomes the increasingly private world of those who explore the cosmos both beyond our universe and within matter, bringing ideas and ways of thinking which are beyond the grasp of most of us. However detached or objective scientists wish or imagine themselves to be, they are learning that the influence of the personal is ineluctable. The understanding of knowledge today and the new technologies mean that data is available as never before. A major consequence is that the motivation of the researcher becomes more significant.

4 Playing God

It came as no surprise to people in the West that Yuri Gagarin, the first cosmonaut to circle the Earth, did not find God on his orbit. There was equally little surprise when an American astronaut, circling the moon alone for the first time, read the opening words of Genesis as he looked down upon the Earth hanging in space. But neither seemed to doubt that the step they were taking from this planet was also taking us into the worlds of creation in the widest sense. In particular, as finite minds began thinking from outside this planet, there was at the same time similar thinking exploring aspects of the body and the mind in new and disturbing ways.

"Playing God" is an easy accusation to make against scientists; some of the discoveries of modern times and their potential for good and ill are so significant that people become afraid. Sometimes the cosmologists, with theories of ever more complex structures at the heart of the universe, make us feel small and even ignorant. It is not easy to grasp the idea of millions of universes racing at incomprehensible speeds through space. It is equally a problem for the "ordinary person" to understand modern medical research, which is so immediate in its outcome and so complex in its conception. They only have to think of, for example, surgery for heart replacement or, even more dramatically, advances in understanding conception and bringing it about. As for people like me,[12] we enthuse with reports on such matters as neurosurgery, stem cells or pharmacology. But like most sufferers I have little confidence that anything useful will be discovered in my lifetime. Science does not console at times of stress any more than religion, but each has a complementary function.

5 Science, religion and authority

Far from being in opposition to each other, as some assume, science and theology in practice share a common interest in the contextualization of theories and hypotheses and here-and-now living. In the religious context the three themes may be for the individual or the community an unfamiliar insight or experience (theory), theological reflection (hypothesis) and faith (living here-and-now).[13] The hypothesis is generated and tested by a different canon of tests from those in a scientific experiment. The outcome is either rejection or positive involvement in the developing life of the religious institution. It may be its worship or social concern. It may be a matter of doctrine. Or it may even lead to the creation of a whole new institution separate from the original.

The pattern is similar for the scientist. He discovers or is struck by an idea that generates a theory that he will test by a controlled experiment. When he has done in his view sufficient work in this area, he will open the work for his peers to examine. They will either confirm, modify or disprove his ideas. If the initial hypothesis is supported by experimental data, these results may be published in support of a larger model or theory. In order for any hypothesis or model to stand, two things have to happen. In the first place, of course, the experiments have to be verified. The attempt to replicate them will be made and, if this confirms the original thesis, there is a second phase: the confirmed work is added to a larger story – which may be embellished, misrepresented or over-stated – and that is then made available for public consumption.[14] The case study on prayer and *in vitro* fertilization, which serves as the opening vignette for this book, offers rich material to illustrate this point.[15]

The process is more complicated if we look at issues on the grand scale such as, for example, evolution. Most scientists, and indeed most thinking people, now consider Darwin's discoveries and hypotheses are sufficient for evolution to take its place as a scientific fact – it no longer has to prove itself. It has been profoundly explored and affirmed. Yet during that time it has also at the popular level become one of the stories that people use to place themselves in the created world, that is, a myth of our age.[16]

Science and religion have held mutual disagreements, at different times, and for a variety of reasons. But only in a few cases is there in fact conflict. Indeed it might not be too recherché to hypothesize that it is the similarity of stance and attitude to the world (or creation) that leads to occasional hostility.[17] The current arguments about stem cell research are an instance of this struggle, but the critical question for both scientists and religious people is not so much a question of method as that of authority. Who can speak for whom with sufficient authority? And when authority is questioned, the issue of truth begins to emerge.[18] This is exactly the same as with religion. Churches and synagogues have sometimes found that mutually respectful conflict is the way in which a body of that sort develops. Where belief is concerned there will be so many factors, ranging from the history of the church to its present position, from old favorites to new forms of worship which are acceptable to all. One outcome of the major essays in this book could be that we shall find ourselves examining more carefully the question of truth.

Authority is a major issue in religion, not least since one function of religion is to bind a society together (Latin root – lig). Although traditionally involved in this task, the churches (I cannot speak for other faiths) have found increasing difficulty with their representative functions and corresponding loss of confidence in those roles. One of these has traditionally been to stand with people in the irrational[19] or spiritual dimensions of their lives and offer interpretation. It is a social function that today is still laid on ministers or clergy in the first instance.

> ...there is inevitably an element of childlike dependency in a relationship to the church, and thus to its representatives, in that to some extent they are being asked to solve the insoluble, to cure the incurable, and make reality go away.[20]

This, too, is not easy in a modern secular society in which there seem to be fewer institutions to bind it together and where in any case such a function may be derided and casual diversity encouraged as a norm. In such a context (for which the term often used is "post-modern") it is easy to claim that religion is a private matter, the personal concern of the individual. The inadequacy of that view is now (as this volume goes to press in 2008) being daily demonstrated in the Middle East and Afghanistan. When that problem is "resolved," we may be reasonably certain it will flare up elsewhere.

When authority is such a core problem, the question of truth also arises unavoidably. And in a media-saturated world there are always critical questions, whether explicit or not, concerning the media's apparent urge to be authoritative and determine truth. We are in the realm of control. This capitalizes on the power of the visual image – "I saw it on television" reinforced by that canard "The camera never lies."

It is difficult to speak at all about religion without being specific. I come from the Jewish-Christian tradition and that is one background to this essay. Those from other traditions might have said something different.[21] At one time the answer to every question that was judged worth asking was to be found in the Bible or Aristotle. It is difficult for us, who have been brought up in an experimental and experiential world, to feel the power that this way of thinking exercised. There are still biblical fundamentalists, but I doubt if there are any pure Aristotelians. The 16th and 17th centuries fundamentally changed what counted as evidence both in the science and religion.

6 "True" and "truth"

During the 19th century it became increasingly difficult to speak about "truth." The tradition of the past seemed not to provide much for the present. Matthew Arnold watched the culture of Christendom flow past him as he stood on Dover Beach.

The Sea of Faith,
was once, too, at the full,
and round earth's shore
like the folds of a bright girdle furl'd.
But now I only hear
its melancholy, long, withdrawing roar,
retreating, to the breath
of the night wind, down the vast edge's drear
And naked shingles of the world.[22]

Neither was the 20th century a time for philosophical grandeur. Scholars turned to analysis. Of truth, for example, J.L. Austin, one of the Oxford philosophers of the 1960s, said, "in vino possibly veritas, but in sober symposium verum" – "When you're drunk you might talk about truth; but when you're sober we can only talk about what is true." That is the style of science: to work on the basis of evidence to find what is true and, at least notionally, to avoid excessive and grand speculations about truth. Yet even as I write scientists in several (I think seven) locations are racing to complete particle accelerators which they claim might expose the nature of matter and answer the question of "the God particle" (Higgs' boson) – ultimately matter itself (although this ultimacy has so often been claimed that the odds must be that it will happen again – Higgs' boson 2?). This seems ambitious. Although the first team to a conclusion will make modest claims, there is no doubt that the mere fact of these groups being engaged in the work at the moment implies competition and eventual kudos and, most importantly, further finance which tends to follow success.

The vignette of the "Brave Bulls" experiment of J.M.R. Delgado, in the mid-1960s, has been described in detail earlier in this volume by Prof. Snyder (Chapter 3, and see accompanying DVD of the film he made of this demonstration experiment in Southern Spain). I believe that Delgado has to answer to what appears at first glance the absence of truth: "When we are sober we can only talk about what is true." Scientific activity is necessarily interactive. Each academic discipline is so vast that few, if any, can specialize in more than one. In addition, the scientist in his exploration of the data also contributes from his imagination and thus helps to help create what is "there," which may then be discovered.[23] I am unsure about the present state of Schrodinger's cat, but he is either there or is not. That is the style of science: to work on the basis of evidence to find what is true and, at least notionally, to avoid excessive claims and grand speculations about the truth.

Delgado's demonstration fits this pattern: the bull, its aggression, ill temper and movement, are given. Delgado's "discovery" of mood control is then shown to the world by the implantation of electrodes in the animal's brain. When the reporter gets hold of the story, the bull turns from "brave" and fierce to docile, the scientist becomes "the unarmed matador" and the whole episode becomes "probably the most spectacular demonstration ever performed of the deliberate modification of animal behavior through the external control of the brain."

Here we see truth dissolving in the face of what may have been a hidden agenda. For this experiment was reported in 1965. And during the 1960s, both in the United States and in Europe, many academic institutions were the loci of the various anti-war and anti-capitalist protests of that period. Because he is "brave" the bull is anthropomorphized into becoming a partner in the experiment, thus to some extent appeasing, or at least recognizing, the anti-vivisectionists; "probably" is a way of covering his back should his report prove to be incorrect. But "deliberate modification" is exactly what people were trying to do with society but were not sure how to do it other than by way of protest, if necessary violence.[24] And "external control of the brain" was both frightening and attractive in the possibilities presented for the authorities to re-establish their control in the face of powerful forces in society.

A theologian listening to this story would hear warning bells ringing, at least on the points mentioned here. In particular, the question arises as to what was the basic motivation for this experiment and its reporting. Because we know this report was inaccurate, the question will be whether there was any collusion to mislead. The answer will have to be that there was, even if it was largely unconscious. In the 1960s, governments and law enforcement officers in the western democracies were concerned about civil rioting. Any new means of controlling civil unrest were likely to be examined. The issue, as we have noted, is one of control. And if anything, concern about that has increased in subsequent years. Whether Delgado was aware of all this is unclear. I understand that he may have received government funding but that may imply nothing. But to my mind his manipulation was an instance of scientific misconduct. There is no "pure" science and this story cannot be read apart from its context. And in an odd way this story paves the way for some well-known episodes of misconduct. Although this may be for a variety of reasons one contributory factor.

7 Multi-tasks in multi-authorship

This may seem a small matter but one that could perhaps encourage in all that trust which is part of scholarship. Articles in scientific journals are in most cases multi-authored. Sometimes so many researchers are listed that we are surprised that they could all fit into one room and even more that they could work there. In other disciplines articles are usually written by an individual, who then takes ultimate responsibility for it and pays due respect to his sources, including colleagues. By contrast, scientists enjoy the multiplicity of authors with no clear indication of who did what. This is usually explained as a consequence of the distinctive way of collaborative working that is a mark of science today.

A highly publicized scandal in medical research occurred in 2006. One of the world's leading pioneers of stem cell research, Hwang Woo-Suk, admitted that he had fabricated results in his research into stem cells. He was internationally renowned for his pioneering work on cloning. It so happens that his research was expected to open up new ways to treat Parkinson's and allied diseases. Not only, therefore, did he deceive his colleagues but he also abused the infirmities of many.

The story of Geoffrey Chamberlain in the United Kingdom is particularly instructive in this regard, and is to date Britain's highest profile case. Professor Chamberlain was editor of *The British Journal of Obstetrics and Gynaecology* and President of the Royal College of Obstetricians. Widely admired and held in affection, he seemed destined for a knighthood. One of his colleagues, Malcolm Pearce, produced a paper which was submitted to the journal and published. Although doubly peer reviewed, the accounts, statistics and patients were all fraudulent. And Chamberlain, without having read the article, put his name to it as a co-author. It became clear that Pearce had gone to great lengths to falsify data and hide it. His fraud was nonetheless discovered and he was both dismissed from his position and reported to the General Medical Council. Soon thereafter, Chamberlain also resigned from the posts of editor and president. The reason why has an almost spiritual tone, as recounted by Richard Smith:

> Why does research misconduct happen? The answer that researchers love is 'pressure to publish', but my preferred answer is 'why wouldn't it happen?' All human activity is associated with misconduct. Indeed misconduct may be easier for scientists because *the system operates on trust* (my italics). Plus scientists may have been victims of their own rhetoric: they have fooled themselves that science is a wholly objective enterprise unsullied by the usual human subjectivity and imperfections. It is not. It is a human activity.[25]

In most academic studies, when the researcher is running a program and writes it up fully, her colleagues may contribute by advising her, suggesting things to do, may be even drafting something. They would be acknowledged in notes. But they are not acting as authors of the article, and the author takes her authority for the piece. Double or triple authorship is quite a complicated thing and it is not as easy as it might appear. But six or more authors are difficult to justify as "authors." Scientists seem to do it this way as an unconsidered convention. For it seems that to be mentioned does not require anything to be specifically attributed to one. There would be more integrity if there was limited responsibility for authorship, which puts the lead writer under that particular accountability to each of her colleagues rather than just providing them with a space to add another article to their name. You can be fairly sure that when someone is announced as having written numerous papers, she has done nothing of the sort but everybody colludes in the notion that she might have done. While we may acknowledge the different styles of working between the sciences and other disciplines and the implications for publication, the argument for change seems strong. Each contributor would not just be named; he or she would have their specific role in the research program defined and described and the authorities for the article would become clearer.

8 Communication in science and religion

We finally consider briefly the impact of modern media. Religion in the 19th and 20th centuries was in ferment, not least as universal education and the media began to play a more influential role. Newspapers flourished because people could read them. Technology enabled reports to be gathered and sent immediately. Other inventions made the mass production of newspapers possible and the rail network distributed them nationally. To this day having seven daily national papers in England still gives the edge of competition between them. It also created larger newspapers with more column inches to fill. This new need to provide a continual onslaught of interesting stories has, at times, encouraged the temptation to fabricate or embellish information, a not uncommon occurrence in reporting (there have been two widely publicized cases of this occurring by reporters at the venerable *The New York Times*, in recent years).

But in terms of disseminating information, this new growth of media was a golden moment in the relationship between science and religion. The railways moved newspapers around the country; and the bishops, who were in the forefront of taking up this new mode of transport, would read the newspapers.[26] When they came to the sermon they might comment on some news of a new discovery. And some deans and bishops were scientifically qualified. Soon specialist science correspondents were appointed, which was the norm until recently. Some were highly qualified and the idea of a science correspondent who had not studied a science was scarcely conceivable. Yet today they sometimes might be journalists who know how to gut a handout, not informed people who seek to interpret science and its discoveries to a discerning audience. The correspondent could also find distinguished scientists who would give him or her advice.[27]

There are now fewer safety nets and stories fall into familiar categories. There is the "wacky" story of the uncontrolled search for that power that may threaten the world; similarly, but not quite the same, there is the paradoxical health story, for example red wine is good for you, as is chocolate. There are scare stories, often manufactured by the media from a small study which has somehow come to the reporter's attention. Finally there are new breakthroughs:

> ...these stories sell the idea that science and indeed the whole empirical world view, is only about tenuous, and new, hotly contested data... often a front page side story will emerge from a press release alone, and the formal academic paper may never appear or appear much later so that the final findings do not even show what the press report claimed it would.[28]

The social costs are dire. But from the perspective of religion more is lost than just social costs. Such behavior and argument demean a series of concepts that many religions claim permeate (or ought to permeate) human life and contribute to its meaning – respect for one's neighbor, concern for the poor and deprived.

Take the word "discovery." The Oxford English Dictionary gives no fewer than 11 meanings. But it begins with the idea of uncovering or exposing what is already there. Scientific discoveries fall into that category, and yet the whole modern scientific enterprise rests on one unprovable and untestable hypothesis, namely: that there is an explanation for everything. Not content with hypotheses about God and the nature of the universe, scientific method finds itself, no doubt with some surprise, modeling a sort of liturgy for common belief. For example, any disaster, whether natural or the result of human error, usually creates a ritual: the "creed" is affirmed – "We shall find out what happened" – accompanied by the ritual mantra and statement of faith that we shall learn whatever lessons may be gleaned. From a formal inquiry a report with recommendations is taken as an explanation – our "faith" is vindicated. After brief, but often intense, media hype, this explanation is accepted and serves as the liturgy for another secular requiem. Here science, religion and media are united.

Scientists and religious people alike are not immune from naivete when it comes to the media. There is a catch, however: if you wish to communicate to a wider audience, you have to use the available media (in the most technical sense of the word – medium: that which occupies the middle). When the media consisted largely of newspapers, both national and local, the press seemed to believe that it had a particular duty to disseminate knowledge.[29] It had a monopoly, there being no competition. An editor might even ask a reporter to produce his notebook. It was not a sophisticated system, but the requirement of accuracy would make a reporter wary of misleading his editor before he misled the public. Today technology has reduced the significance of a medium: a reporter types the material straight into the document where, subject to examination by lawyers, it is usually published as it stands.

The new mass media now largely shape the world. Science and religion are not immune from this. It is mostly covert, but from time to time it becomes explicit. For example, newsreaders may refer to the studio base of the news program as "the show." News, as news, is being increasingly confined to news channels, which are not watched by the bulk of the population. The future of democracy is at stake, for in the long term it cannot afford an ignorant electorate. Part of the problem is, however, that concepts are increasingly difficult for the non-specialist to grasp, whether they be stem cells or quantum physics. Another is the introduction of 24-hour news channels. The rush to judgments and incomplete stories give undue prominence to minor issues. On the other hand, mistakes can be quickly corrected in a rolling bulletin.

For an example of the impact of media intrusion, consider the film Witness (1984). It is set in the Amish community which for religious reasons isolates itself so far as is possible from the surrounding American culture. They are specifically not allowed to watch television. Despite this, several people went to watch the filming. One boy had cut out a picture of Harrison Ford and pinned it on his bedroom wall.

> "This doesn't matter to us," his mother said. "Someone told us he was in Star Wars, but that doesn't mean anything to us". The last time a similar conclusion was drawn was the executive director of the American Association of Blacksmiths' remark that he had read about the automobile but he was convinced it would have no consequences for the future of his organisation.[30]

In order to begin communication, therefore, we have to accept that television has become "the command center of the new epistemology," namely bodies like hospitals and churches are weaker as symbol bearing institutions. If any new scientific discovery, or even new theological insight, is to be brought into the public arena, it will undergo some form of editing by the media. It is ultimately their judgment whether to risk publication. For the authors' part it is a matter of the extent to which they are to be trusted (to which the answer is often "very little").

As I write, however, we are beginning to wonder whether television itself is now becoming old and that more communication is likely to be through the Internet screen and the iPod or iPhone. To be on-line is becoming essential and the web is itself an impressive achievement of applied science. But it is also something of general concern. Many are worried about the abuse of the Internet, and rightly so. But a more serious concern is the way in which we lose our capacity for discernment. Then truth in any subject becomes a matter of opinion only, there being no way of evaluating the various contributions to the world-wide net. We are entering new realms of knowledge overload. But without the current old skills we shall have no critical base from which to assess the value of any presentation. Already the lawyers are telling us that we have very little control over what is said on the Internet. The moral questions are acute.

9 Conclusion

The name "Dolly" is not calculated to terrify. Yet the sheep by that name has caused some anxiety and provoked discomfort at the boundary between science and religion. For Dolly (1996–2003) was a cloned sheep. And if we can create a sheep, the argument goes, why not human beings? It is reminiscent of the film O Lucky Man, which was produced in the 1960s. In it the hapless anti-hero, played by Malcolm McDowell, visits a research station and that night as he looks around he sees a man lying on the bed. He turns back the sheet to find that the head is sewn onto a sheep's body. It has only taken about 30 years for what was inconceivable to become almost a potential norm. The strict parallel, however, is not between Dolly and that half-and-half creature: the controversy is over stem cells and using a small amount of human DNA to create active cells which could possibly heal the most serious brain disorders. Here, if anywhere, is the currently important encounter between religion and science.

There is no one view on the part of people of faith. The dispute is not about the use of these cells, but their acquisition from embryos. In America it is not just a religious question: it is also highly political. The issue may be more easily handled in Europe, although there is little agreement on the social, political and religious significance of an embryo. We could end up in the odd position that scientists are ready for clinical trials but the politicians, and maybe even some religious leaders, will inhibit them. There is widespread ambivalence: on the one hand it will do great good, so let us have it. On the other hand, can we trust scientists not to go too far and create a catastrophe? Lack of interest rather than the urgency of an interface between religion and science is today's culture. For example, the international Human Genome Project was completed early and is already a vital tool for scientists working in that field. However, this extraordinary achievement was not afforded the major publicity it deserved. Perhaps that was because the scientists involved could see ordinary people becoming scared at the potential implications. A useful distinction here is between learning about some scientific discovery and recognizing its implications. Even more pressing is some conclusion to the debate on climate change, in which lay people's judgment and perception in the light of reported scientific analyses will be critical. We are again facing issues of truth and authority.

Perhaps during the past 200 years or so any tension between religion and science has been so addressed that they can be recognized more as allies than as enemies. For both now are old and still integral parts of a civilized society. Secularization has not destroyed religious faith; rationality alone seems to be inadequate to deal with the complexities of the individual psyche and the social context in which we live. Irrationality still rears its head: Smith's comment on research misconduct (noted above) made that clear. Irrational behavior is not confined to the arts or religion: it permeates human life and some way every system has to cope with it. And all of us have to be careful with projection; it is easy to demonize the "other" without examining oneself. This is something we have learned from the "new" behavioral sciences. But, as we have seen, in the light of the development and expansion of the media there is now always that third factor to be taken into account. And here, so far as I can see, both science and religion find themselves in the same situation: their messages are potentially significant but experienced largely as incomprehensible. Yet if they work with rather than against the media, they may find that genuine communication takes place. The major hope of development and building of confidence between anxious people of faith and adventurous scientists lies in openness and from all parties an appropriate humility in the face of the vastness of contemporary scientific discoveries. For that, as we have seen, society needs a reasonable, if not perfect, media. Perhaps the best we can hope for – and it is not a bad hope – is that truth will prevail at least a little bit.

Acknowledgment

I am grateful to James Wilkinson, for 25 years Science Correspondent of the BBC, for his interest and help.

Pat Shipman, Ph.D.
Adjunct Professor, Department of Anthropology
The Pennsylvania State University
University Park, PA, USA

Dr. Pat Shipman has led a diverse career path, reflecting two distinct threads of enduring interest: writing and science. Dr. Shipman spent 20 years as a practicing scientist, earning an international reputation for her work on early human ancestors and their ecological adaptations and niches. She proposed a once-controversial theory – now generally accepted – that our early ancestors underwent a dietary shift from plant-eating to meat-eating, by developing adaptations that were initially useful for scavenging rather than hunting. Shipman has analyzed and sometimes excavated fossil materials from various parts of Africa, to glean evidence on the adaptations and fundamental biological traits of our early ancestors, and she developed innovative techniques for analyzing fossilized assemblages of bones and teeth. Dr. Shipman also applied these techniques to materials from Europe and the Americas. As Dr. Shipman investigated human origins and adaptations, she became fascinated by the history of paleoanthropology and the development of ideas: who proposed which theories and why? How do interactions with mentors or contemporaries shape the course of the research? Why does a particular individual formulate the theories he or she does? And how do gender, ethnicity, social standing, and networks of friendships influence the formation and acceptance of scientific theories?

Since resigning her academic position in 1990, Dr. Shipman has developed a second career as a biographer and as a writer of science for non-scientists. Her career trajectory has afforded her an unusual insider's perspective on the human aspects of science. Shipman believes that there is an urgent need to remedy the gaps in the public understanding of science through lively, vibrant, and accessible writing. To this end, she has authored more than 50 scholarly articles and over 100 articles for the general public in such diverse venues as *Discover*, *Natural History*, *New Scientist*, *Anthroquest*, and *American Scientist*. Dr. Shipman has written two technical and eight popular books (*Taking Wing* and *Wisdom of the Bones*), with her writing earning several major literary prizes. She has been selected as the A. Dixon and Betty F. Johnson Lecturer on the Communication of Science, as a Fellow of the American Association for the Advancement of Science, and as a Fellow of the Royal Geographical Society in the United Kingdom. In 2005, Dr. Shipman was awarded the Leighton Wilkie Prize by the Stone Age Institute and Indiana University, for her contributions to Paleoanthropology. Dr. Shipman's latest book, currently in press with Harper Collins, is *Femme Fatale: Love, Lives, and the Tragic Life of Mata Hari*.

CHAPTER

10

Science Meets Fundamentalist Religion

Pat Shipman

Some 30 years ago, when I was a newly minted Ph.D. in anthropology, I went to visit my sister, a librarian in a small rural high school. When the biology teacher heard I was coming, she asked if I could speak to her class about the fossil site in Africa that I had analyzed for my thesis. I was flattered and anxious to speak to the younger generation about my research into human origins, so I happily agreed.

"But," the biology teacher cautioned, "you can't mention the word 'evolution.'"

"*What*?" I asked, incredulous.

"This is a very conservative district," she replied. "I don't teach openly about evolution." She was afraid of losing her job. She added, "The best thing would be if you can get the students interested enough that they ask you about evolution; then you can answer as fully as you like."

I wasn't afraid of losing my job, or even my lecture fee, because I had neither. I did my best to light the fire of curiosity in the minds of those kids. I wish I had that chance again now, because I don't think my inexperienced best was very good. I didn't get many questions that let me fill in the enormous gap in their knowledge of science. This is a single incident, from years ago, and might not signify much if similar scenarios were not being enacted, more overtly, throughout the United States and much of the world on a daily basis.

1 The Dover trial

One of the best-documented examples of this phenomenon occurred in Dover, Pennsylvania, a town not far from where I live. In 2005, 11 Dover parents, with the assistance of the American Civil Liberties Union and Americans United for Separation of Church and State, brought a lawsuit to stop the reading of a disclaimer about evolutionary theory and to stop the teaching of Intelligent Design (ID) in science classes.

The lawsuit was provoked by changes made in 2004 and 2005 by the school board which incorporated weakened the teaching of evolution in high school biology classes and recommended a book promoting ID. The new mandate for teaching biology read in part:

> Students will be made aware of gaps/problems in Darwin's theory and of other theories of evolution including, but not limited to, intelligent design. The origins of life will not be taught.[1]

At least some members of the Dover School board had a clearly religious agenda. Chairman of the school board Alan Bonsell wanted equal amounts of time spent teaching evolution and creationism and hoped to reinstitute prayer in school.[2] Chairman of the curriculum committee, William Buckingham objected to buying the new edition of the textbook *Biology* because it was "laced with Darwinism."[3] At a board meeting, Buckingham said, "Two thousand years ago somebody died on the cross for us. Shouldn't we have the courage to stand up for him?"[4] He also asserted, "It's inexcusable to teach from a textbook that says man descended from monkeys."[5]

When the head science teacher expressed her concern that teaching ID might violate the separation of Church and State, Buckingham replied, "Nowhere in the Constitution does it call for a separation of Church and State."[6] On this point, Buckingham is flatly wrong, if the Bill of Rights is considered part of the Constitution. The First Amendment to the Constitution reads,

> Congress shall make no law respecting an establishment of religion, or prohibiting the free exercise thereof; or abridging the freedom of speech, or of the press; or the right of the people peaceably to assemble, and to petition the Government for a redress of grievances.[7]

In law, the first two phrases of this amendment are known together as the Establishment Clause. In 1947, the Establishment Clause was interpreted by the Supreme Court in Everson v. the Board of Education:

> The establishment of religion clause means at least this: Neither a state nor the federal government may set up a church. Neither can pass laws that aid one religion, aid all religions, or prefer one religion over another. Neither can force a person to go to or to remain away from church against his will or force him to profess a belief or disbelief in any religion... Neither a state nor the federal government may, openly

> or secretly, participate in the affairs of any religious organizations or groups and vice versa. In the words of Jefferson, the clause against establishment of religion by law was intended to erect 'a wall of separation between church and state.'[8]

Urged on by Buckingham and Bonsell, the school board of Dover High School insisted that the teachers read a disclaimer about evolution and ID to each biology class. This disclaimer said:

> The Pennsylvania Academic Standards require each student to learn about Darwin's Theory of Evolution and eventually to take a standardized test of which evolution is a part.
>
> Because Darwin's Theory is a theory, it is still being tested as new evidence is discovered. The Theory is not a fact. Gaps in the Theory exist for which there is no evidence. A theory is defined as a well-tested explanation that unifies a broad range of observations.
>
> Intelligent design is an explanation of the origin of life that differs from Darwin's view. The reference book *Of Pandas and People* is available for students to see if they would like to explore this view in an effort to gain an understanding of what intelligent design actually involves.
>
> With any theory, students are encouraged to keep an open mind. The School leaves the discussion of the Origins of Life to individual students and their families. As a Standards-driven district, class instruction focuses upon preparing students to achieve proficiency on Standards-based assessments.[9]

As was pointed out during trial testimony, the disclaimer singled out evolution as an apparently "half baked" theory,[10] in distinction to every other scientific theory presented to the students. Brian Alters, an expert in science education, also emphasized that reading the disclaimer required teachers to disregard findings of the scientific community and to teach in ways contrary to the recommendations of the National Association of Biology Teachers.[11] Alters phrased the implicit message of the disclaimer as: "We have to teach you this stuff, you know... We'd rather not do it, but Pennsylvania Academic standards ... require students to do this."[12]

During the 2005 trial, biology teacher Jennifer Miller testified that intimidation tactics were used, such as threats of firing teachers who objected to the changes and the ridiculing of students who opted out of hearing the disclaimer. The book cited in the disclaimer, *Of Pandas and People*, was not a reference book akin to a dictionary or encyclopedia but was blatantly religious in nature. First, the book was written by three avowed creation scientists.[13] Second, it was published by an organization that identified itself to the Internal Revenue Service (IRS) as a Christian organization.[14] Third, and perhaps most tellingly, in early drafts the book defined "creation science" in terms identical to those used to define ID in the final version. Those early drafts were written prior to the Supreme Court decision of 1987, Edwards v. Aguillard, in which creation science was ruled to be "manifestly religious,"[15] meaning creation science could not be legally taught in public schools in the United States. Subsequent to this ruling, *Of Pandas and People* was revised and the authors replaced the word "creation" or its cognate in 150 different places in the text with the phrase "Intelligent Design," which appeared in the published version.[16] Fourth, the book was recommended to Buckingham by the Thomas More Law Center, a Christian-oriented law firm from which he sought legal advice and which provided the defense for the Dover School Board in the lawsuit. Finally, copies of the book were donated to the Dover school by a religious group – the Buckingham's church, the Harmony Grove Community Church[17] – a point that both Buckingham and Bonsell attempted to conceal.[18]

Seven biology teachers in Dover refused to read the disclaimer to their classes and risked their jobs by writing a powerful letter to the superintendent of schools, Dr. Richard Nilsen. The letter read, in part:

> 'INTELLIGENT DESIGN' IS NOT SCIENCE. 'INTELLIGENT DESIGN' IS NOT BIOLOGY. 'INTELLIGENT DESIGN' IS NOT AN ACCEPTED SCIENTIFIC THEORY.[19]

The teachers argued that treating ID as a scientific theory on a par with evolution would constitute unethical conduct on their part, writing,

> Central to the Teaching Act [The Pennsylvania Code of Professional Practice and Conduct for Educators] and our ethical obligations is the solemn responsibility to teach the truth. Section 235.10 guides our relationships with students and provides that the professional educator may not knowingly and intentionally misrepresent the subject matter or curriculum.[20]

This is one of the most striking and powerful arguments to be made by pro-evolutionists: that it is morally and professionally unethical either to misrepresent evolutionary theory as a shaky hypothesis or to represent ID as a scientifically credible one.

The Dover lawsuit became a landmark case on the merits of ID. Judge John E. Jones III found ID to be a manifestly religious doctrine that failed to meet the criteria of science. He wrote in his opinion:

> ...ID is not science ... ID fails on three different levels, any one of which is sufficient to preclude a determination that ID is science. They are:
>
> (1) ID violates the centuries-old ground rule of science by invoking and permitting supernatural causation; (2) the argument of irreducible

> complexity, central to ID, employs the same flawed and illogical dualism that doomed creation science in the 1980s; and (3) ID's negative attacks on evolution have been refuted by the scientific community... [I]t is additionally important to note that ID has failed to gain acceptance in the scientific community, it has not generated peer-reviewed publications, nor has it been the subject of testing and research.[21]

In coming to his conclusion, Jones referred to the legal description of science from a previous Supreme Court case:

1 [Science] is guided by natural law.
2 It has to be explanatory by reference to natural law.
3 It is testable against the empirical world.
4 Its conclusions are tentative, that is are not necessarily the final word.
5 It is falsifiable.[22]

In his view, ID did not meet these criteria. He wrote in his judgment,

> [S]ince ID is not science, the conclusion is inescapable that the only real effect of the ID Policy is the advancement of religion... The effect of the Defendants' actions in adopting the curriculum change was to impose a religious view of biological origins into the biology course in violation of the Establishment Clause.[23]

The issue of teaching ID in public schools in Pennsylvania was settled definitively by Jones' judgment, which also served to alert ID movements in other statements that a new legal tactic would be required if their intention of putting ID into the classroom were to be successful.

A key ethical issue was raised by the Dover case. Various members of the school board attempted to impose their religious beliefs upon others and made significant efforts to conceal their religious biases and remarks. The behavior of Bonsell and Buckingham was so blatantly duplicitous that the judge referred to their "repetitious, untruthful testimony" and their "flagrant and insulting falsehoods to the Court."[24]

Regardless of the legal consequences of such actions, is it ethically correct or moral to lie in order to promote a religious belief? Is it ethically excusable to impose one's own religious beliefs upon others at all? While I make no claims to be an ethicist or a scholar of ethics, my considered opinion is that sacrificing honesty and respect for others in the hopes of spreading one's own moral beliefs is a dangerous and possibly immoral tactic.

2 The El Tejon case

Similar issues were raised in a more recent case, which involved the proposal to teach a course entitled Philosophy of Intelligent Design in El Tejon California. According to the course description, the class would

> take a close look at evolution as a theory and will discuss the scientific, biological, and Biblical aspects that suggest why Darwin's philosophy is not rock solid. This class will discuss Intelligent Design as an alternative response to evolution. Topics that will be covered are the age of the earth, a world wide flood, dinosaurs, pre-human fossils, dating methods, DNA, radioisotopes, and geological evidence. Physical and chemical evidence will be presented suggesting the earth is thousands of years old, not billions. The class will include lecture discussions, guest speakers, and videos. The class grade will be based on a position paper in which students will support or refute the theory of evolution.[25]

A lawsuit was instituted against the El Tejon school board, by both parents in the school district and by Americans United for Separation of Church and State, to prevent the teaching of this course. Advice was offered to the school board by Casey Luskin, an attorney representing the Discovery Institute, a prominent ID think tank with the stated aim: "To replace materialistic explanations with the theistic understanding that nature and human beings are created by God."[26] Fellows of the Discovery Institute include some of the authors of the book *Of Pandas and People*, which figured prominently in the Dover trial. Luskin told the El Tejon school board:

> From what I can tell, this course was originally formulated as if it would promote young earth or Biblical creationism as scientific fact. Although I understand that the course has since been reformulated to remove the creationist material, a course description was sent out to students around December 1st which described this course as promoting young earth or Biblical creationism as scientific fact. This is very concerning because courts have made it clear … that young earth creationism is unconstitutional to teach as fact in public schools.
>
> Intelligent design is very different from young earth creationism. We at the Discovery Institute believe that intelligent design is constitutional to teach as a science… I want you to know that we support your efforts to present different views about biological origins in this philosophy course. We also applaud your efforts to remove the legally problematic creationist materials from the course. But the fact of the matter is that even if this course has been changed and improved, its … originally having been formulated to promote Biblical creationism as scientific

> fact makes this case legally problematic. Unless you get a very sympathetic judge, this course will be struck down as unconstitutional...
> [G]iven the history of this course, this course threatens to become a dangerous legal precedent which could threaten the teaching of intelligent design on the national level. The young earth creationist history of this course places it on extremely shaky legal ground...
> [T]he only remedy at this point to avoid creating a dangerous legal precedent is to simply cancel the course.[27]

He advised them to regroup, seek council more widely, and offer the course again next year in a more appropriate form. Luskin's concern was to dissociate ID from the creationist movement and to prevent a lawsuit being brought that would result in another judgment against the teaching of ID in the public schools. He encouraged them to circumvent the law yet still inject ID into the curriculum at a future date.

The settlement reached by the School District with Americans United for the Separation of Church and State on January 17, 2006, prohibited the School District from offering the course "entitled 'Philosophy of Design' or 'Philosophy of Intelligent Design,' or any other course that promotes or endorses creationism, creation science, or ID" in subsequent years.[28]

Was Luskin advising compliance with the law or was he suggesting the El Tejon school board use clever strategies to avoid prosecution while violating the intent of the law?

3 Broader difficulties in teaching evolution

These American towns are not the only localities to experience overt or covert struggles between the teaching of science and religious beliefs of the populace. Fully one-third of the science teachers polled by the National Science Teachers Association feel pressured to include ID, creationism, or other "nonscientific alternatives" in their science classrooms.[29] Others are intimidated by the threat of parental complaints and simply skip or shortchange material dealing with evolution in their classes. In 43 of the 51 states (including Puerto Rico), attempts have been made to put ID into the science curriculum in public schools. To try to counteract this pressure from fundamentalists, formal statements or resolutions in support of evolutionary theory have been made by at least 61 scientific or scholarly organizations and 42 educators' organizations.[30]

Scientific literacy in the United States has reached a disturbingly low level.[31] By the end of high school, our students rank 19th in scientific literacy and mathematics among students from 21 nations, according to a National Science Board comparison based on testing done in 1995. The number of college students continuing their education to pursue science majors is also disturbingly low, only about 15 percent.[32] Scientific literacy will fall further if public high schools fail to teach the distinction between science and religion to our students.

Despite the conflict between science and religion perceived by leaders of fundamentalist religious groups, many prominent religious authorities do not agree and have issued formal statements endorsing evolutionary science, such as Pope John Paul II, the Archbishop of Canterbury, the Bishop of Oxford, a group of 188 Wisconsin Clergy, the American Jewish Congress, the American Scientific Affiliation, the Center for Theology and the Natural Sciences, the Central Conference of American Rabbis, the Episcopal Bishop of Atlanta, the General Assembly of the Presbyterian Church (USA), the General Convention of the Episcopal Church, the Lexington Alliance of Religious Leaders, the Lutheran World Federation, the Roman Catholic Church, the Unitarian Universalist Association, the United Church Board for Homeland Ministries, the United Methodist Church, the United Presbyterian Church in the USA, among others.[33]

Similar struggles in educational and research policies are beginning to occur in many parts of the world. Britons who had watched with detached amusement as the United States wrestled with attempts to infiltrate ID or creation science into public school curricula are now appalled to find the same battle on their front doorstep. On September 18, 2006, packets containing two DVDs and a manual were sent by a limited company called Truth in Science to the head of science in every secondary school in the United Kingdom. The DVDs feature many of the fellows of the US-based Discovery Institute. Dozens of schools have started using these packets in their teaching. British educators, scientists, and church leaders including the Archbishop of Oxford have reacted with vibrant protests. In the words of Jim Knight, a minister in the Department for Education and Skills, "Neither intelligent design nor creationism are recognised [sic] scientific theories and they are not included in the science curriculum."[34]

Similar attempts to incorporate ID into school curricula have been reported around the world. In Holland, Maria van der Hoeven, Dutch minister of education, culture, and science, suggested introducing ID into Dutch classrooms in hopes of stimulating an academic debate. Instead, she provoked outrage from members of Parliament and the editors of the *NRC Handelsblad*, a major Dutch newspaper.[35]

In Poland, the deputy education minister, Mirosaw Orzechowski, went farther, declaring, "The theory of evolution is a lie … It is an error we have legalized as a common truth."[36] His father, a prominent and very right-wing politician in the European Parliament, is lobbying for the inclusion of creationism in the Polish biology curriculum.

In Kenya, Bishop Boniface Adoyo, the chairman of the 6-million-strong Evangelical Alliance [of Pentecostal churches] of Kenya, urged the National Museums of Kenya to play down the importance of the unparalleled collection of Kenyan fossils demonstrating human evolution. Such exhibits, Adoyo argues, are "creating a big weapon against Christians that's killing our faith. When children go to museums they'll start believing we evolved from these apes." He added, "Our doctrine is not that we evolved from apes and we have grave concerns that the museum wants to enhance the prominence of something presented as fact which is just one theory."[37]

Fundamentalist leaders exercise a great deal of influence over education policy in many Muslim countries. A literal interpretation of the Koran can pose insurmountable barriers to scientific research and education, as in Afghanistan under the Taliban.

Turkey has long been one of the most modern and secular states with a large Muslim population, but creationism and ID have been gaining power there since 1989. In September of 2006, a free, lavishly illustrated book, *Atlas of Creation*, and packets of ID material were sent to many science educators in hopes they would be adopted as part of the curriculum – and sometimes were. The author of the book, Harun Yahya, is the pseudonym of Adnan Oktar, the founder of a creationist group called Bilim Ara ırma Vakfı (the Science Research Foundation), and the author of other works as *The Evolution Deceit*, *The Religion of Darwinism*, and *The Holocaust Hoax*.[38] Yahya's articles and books repeatedly assert that evolutionary theory is a deliberate deceit, a form of nature worship.

> Darwinists' present-day beliefs are just as odd and irrational as those of people who once worshipped crocodiles. Darwinists regard chance and inanimate, unconscious atoms as a creative force, and are as devoted to that belief as if to a religion...
>
> It is unacceptable for students to be taught Darwinist shamanistic religion under the name of science. What they need are courses in biology that have been purged of this shamanistic religion.[39]

Joining Yahya in this regard is prominent Turkish intellectual, Mustafa Akyol. Aykol expressed his reasons for advocating ID in an on-line debate with Nicholas Matzke of the National Center for Science Education, the foremost American institution defending the teaching of evolution in the United States. Aykol wrote:

> There is no problem between Islam and "evolution" per se, but the meaning of the latter is open to interpretation. If evolution means that there is continuous change in nature and that species adapt to their environments, I think that's a scientific statement and it has no contradiction with the Muslim faith. However, by the term evolution, modern Darwinians mean something more. They argue that life on Earth is the product of the blind forces of nature and that there is no Creator worth speaking about... Now I don't think this openly atheistic idea

> is compatible with Islam. And it is not compatible with the scientific evidence, either. It is a philosophical presupposition that is imposed onto the scientific evidence. Intelligent Design theorists have unveiled this crucial fact and that's why they have been receiving so much reaction.
>
> It is more of a debate between theism (the belief that there is a God) and atheism on a scientific level.[40]

In the same way that being labeled a religious movement taints ID in a legal context, so does being labeled an atheistic movement taints science in a religious context. Both groups argue that these labels are incorrect.

4 Science and the media

While educators face ongoing difficulties in the teaching of science, newspapers, magazines, television programs, and books in the United States and much of Europe cover evolutionary discoveries with relatively little censoring, though sometimes with less than desirable accuracy in content. Is open discussion of evolutionary topics a sufficient counterbalance to political movements fostered by fundamentalist religions?

I believe it is not, for the simple reason that the very first step in the dissemination of knowledge – the step that comes long before the free publication of words and opinions – is to *speak* openly and freely. If teachers are not permitted to teach their subjects in accordance with their understanding of the subject (as certified by state-approved credentials and testing methods), they are being denied one of their most valuable and important liberties. And it is this step of talking about theories that conflict with the religious beliefs of particular sects or groups that is being censored and inhibited. The current swing toward fundamentalism is a dangerous movement indeed.

Another problem is that the media does not do its full part in educating the public about science. Media presentations rarely focus on the underlying methodology of science, the lengthy process of hypothesis testing through the gathering of data and observations. Instead of dwelling on the push and pull of scientific inquiry and the testing process that any good new hypothesis provokes, media stories often focus on startling discoveries and overturned theories. A good (well-crafted) hypothesis that is eventually disproven tends to be treated as a disgrace or as reflecting a lack of insight on the part of the hypothesizer. Instead, the media needs to celebrate the investigation that a good hypothesis spurs and the vitality that conflicting hypotheses bring to a field. Being wrong in science means a researcher has successfully narrowed the field of possible rights, not that the researcher has been discredited.

5 Science and morality

In many parts of the world, education is being strongly influenced by fundamentalist and highly conservative religious movements that perceive their beliefs to be in direct conflict with scientific theories, most especially evolution. The fact that evolutionary theory is so often singled out is an indication of its far-reaching and deep-seated explanatory power. Evolutionary theory, broadly defined, accounts for the existence and nature of all of the living creatures on the planet. Because of its vital connections to other sciences, biological science has become highly politicized, both in its conduct and its reporting. In my view, this turn of events is nothing short of tragic.

First, I speak as an American. The United States of America was founded explicitly on the basis of freedom of religion and the freedom to worship (or not to worship) in the religion of one's choice. In this, the United States is an unusual nation, since there are strong, historic ties between church and state in many parts of the world. Thus, in the United States, the state or federally mandated inclusion of manifestly religious content in public schools violates one of the oldest and most honored principles of this nation. Because so many settlers in the New World fled the Old World because of religious persecution, the right to religious expression and to be free from government coercion to behave in accordance with religious beliefs in general or particular is one of Americans' most precious rights. To abrogate that right is surely tragic.

Second, as the science teachers of Dover courageously pointed out, there is an enormous moral issue at stake. It is the moral duty of educators to teach the truth to their students as that truth is known and presented by reputable authorities in the field. This principle does not mean that the truth taught in schools is unchangeable and ever-lasting. Even such a simple thing as the number of states in the United States changes, not to mention detailed interpretations of science, history, literature, philosophy, and numerous other topics. But a teacher's deep obligation to teach to the best of his or her knowledge is very real and very essential. Asking a teacher to do less than his or her best – to lie to students or to tell less than the full truth to them – is surely downright obscene. Teaching is an important and esteemed profession; we cannot demand that its practitioners deceive, dissemble, or prevaricate.

Third, there are serious moral issues at stake among the fundamentalists trying to interject ID into the public schools. Their actions are prideful, intolerant of other points of view, sometimes dishonest, and, to a certain degree, arrogant. Pride, dishonesty, and arrogance are not attitudes or behaviors endorsed by the religions they are attempted to promulgate.

It cannot reasonably be argued that those who would keep religion out of science classes are engaging in reciprocal pride, intolerance, dishonesty, and arrogance. Teaching religion, especially in the context of comparative religion, is entirely legal in public schools. There is (to my knowledge) no political movement that is attempting to prevent the teaching of religion in public schools, as long as one religion is not endorsed over the others, nor is there a movement to interject the teaching of evolutionary science into religion classes.

Finally, both the scientific community and the media have fallen short of fulfilling their compelling obligations to disseminate information that has been gained through scientific research as widely as possible. Most scientific research today is conducted on publicly provided funds. Most journalistic reporting of science is, similarly, paid for by the public. There can be no excuse for using public monies to gain knowledge and then failing to share it as widely and accurately as possible with the public.

To implement this principle fully, researchers need to upgrade their skills in communicating with general audiences and need to take seriously their obligation to do so. Spouting undigested technical jargon at a reporter does not convey the importance and excitement of scientific research in any meaningful way.

Likewise, journalists, reporters, and others in the media need to revise their standards for a "good story" and begin to play their part in educating the public to more subtle nuances. Science reporting today tends to focus on extremes: the biggest, the smallest, the oldest, the newest, or the one-shot discovery that "overturns all previous theories." If spreading knowledge is the key role of educators, then summarizing and publicizing that knowledge is the key role of the media. Reportage needs to include education. Taking this moral obligation seriously may require media professionals to educate themselves more fully than is customary at present. What's more, the media must begin to credit the public with sufficient intelligence to understand and enjoy science reporting that is more complex than what they usually receive today.

Whether one honors God and science, or God OR science, these moral imperatives call us to attention.

Ruth J. Katz, J.D., M.P.H.
Dean, School of Public Health and Health Sciences
The George Washington University
Washington, DC, USA

Ruth J. Katz is the Dean and Walter G. Ross Professor of Health Policy of the School of Public Health and Health Services at The George Washington University, the only school of public health in the nation's capitol. A *magna cum laude* graduate of the University of Pennsylvania, Prof. Katz holds a law degree from Emory University and a Master of Public Health from Harvard.

Her current community service activities include membership on the boards of Emory University, the CDC Foundation, and NARAL Pro-Choice America. She also serves on the National Advisory Committee for the Robert Wood Johnson Foundation Health and Society Program, and the Finalist Selection Committee of the Harry S. Truman Scholarship Foundation. Prior to joining The George Washington University in 2003, Prof. Katz was Associate Dean for Administration at the Yale University School of Medicine, where she also held appointments in the departments of Internal Medicine (General Medicine) and Epidemiology and Public Health as an Assistant Professor.

From 1982 to 1995, Prof. Katz served as Counsel to the Subcommittee on Health and the Environment (chaired by Congressman Henry A. Waxman) of the Committee on Energy and Commerce, in the US House of Representatives. In that capacity, she was instrumental in the enactment of many major pieces of legislation affecting the public health and federal health care financing. Prof. Katz has received several awards recognizing her efforts, including those from the American Academy of Pediatrics, the American Medical Women's Association, and the Association of Community Health Centers.

CHAPTER

11

Uneasy Alliance: The Intersection of Government Science, Politics and Policymaking

Ruth J. Katz

The promise of fetal tissue research has intrigued biomedical researchers for decades because the tissue grows rapidly and can adapt to new environments. In the late 20th century, scientists hoped that it could eventually be transplanted into adults to treat Parkinson's disease, diabetes and other disorders.

But because much of the tissue available for study originates with induced abortions, fetal tissue transplantation research has also been an ideological battleground, especially after the Supreme Court's *Roe v. Wade* decision legalized abortion in 1972.[1] Over the years, Congress has imposed, then lifted, a series of federal funding moratoriums as various advisory groups weighed in with recommendations to guide the work.

Bernadine Healy was a member of one such body, the Human Fetal Tissue Transplantation Research Panel of the National Institutes of Health (NIH), convened in 1988 to examine ethical, legal and scientific questions. A cardiologist by training, Healy headed the Cleveland Clinic Foundation's Research Institute at the time. In her capacity as a private citizen and a scientist, she voted with the panel's majority to say that NIH should support research on the transplantation of fetal tissue.

A few years later, Healy became director of the NIH under the administration of the first President George Bush, which opposed the work. Asked at her congressional confirmation hearing about her stance on the issue, Healy said, "Sometimes science needs to take a 'time out' when its goals collide with the moral concerns of the society."[2]

When the NIH's programs came up for reauthorization in April 1991, Congressman Henry Waxman, chair of the US House of Representatives

Subcommittee on Health and the Environment, had this exchange with Healy:[2]

"Do you know of any scientific reasons why fetal tissue transplantation research should not proceed?" Waxman asked. "Is there a scientific reason apart from the social debate?"

"There are two components," said Healy. "One is the science. NIH reviewed the science.... The conclusion was that the science itself was exciting and promising. It was not a dream fulfilled, Mr. Waxman, but it was something that offered a great promise.

"The social debate, however, could not be ignored. Science must move in the context of society. Science must move in response to the public concerns and public issues.... That particular policy issue involved ethics, religion and moral values that transcended NIH."

The contrast between Healy's scientific support of fetal tissue transplantation research as a private citizen, and her defense of the administration's federal funding moratorium as a government policymaker, was harshly criticized in some quarters. But Healy had not altered her scientific views. Rather, in her role as NIH director she was no longer acting solely as an arbiter of scientific facts. She was also speaking on behalf of the policymaking decisions of the government officials she now represented.

The evolution of Healy's public position raises important questions about the interplay among science, policymaking and politics. Often, their alliance is uneasy, although certainly not always. Many policies can be built on a solid foundation of science without triggering a partisan or ideological response. It is the ones that raise delicate issues of morality, involve significant resource allocations, or involve powerful interests with a vested stake in the outcome that invariably generate controversy.

This chapter reviews case studies from five White House administrations to illustrate how science is used, and sometimes misused, in government decision-making and the options scientists have in responding to those decisions.

1 Science and politics: A continuum of engagement

The vast American scientific enterprise influences virtually every aspect of our lives. With a budget of some $29 billion in FY 2007,[3] the NIH is by far the nation's largest public funder of biomedical research, but it represents only a modest portion of the total funds spent on scientific endeavors. Space exploration, environmental science, energy, infrastructure and other technological avenues of development also cost billions in taxpayer dollars.

In a modern, science-based society, progress builds on a foundation of facts and data and rigorous, peer-reviewed analyzes of what they tell us. Good science, defined broadly for the purposes of this chapter as the process by which we systematically accumulate verifiable knowledge, is essential to informed policy. We can not decide how to regulate pollution until we identify the contaminants in air, water and soil, and understand their sources and health implications. We can not consider strategies for preventing disease until we understand its cause. We can not protect endangered species, make approval decisions about pharmaceuticals, or educate teenagers about safe sex practices until we know what works, and for whom.

But it is naïve to believe that scientific findings are the sole determinant of policy. Much of the funding, direction and use of American science is determined by the federal government and the political biases of the dominant party invariably influence the decisions that get made. So, too, does the obligation to consider the needs of many constituencies with manifold social, economic and ideological concerns.

Whether they are employed at the NIH, the Centers for Disease Control and Prevention (CDC), the National Aeronautics and Space Administration (NASA), or elsewhere, most government scientists learn to recognize this. More so than scientists in academic or industry settings, they may eventually discover that their work is being used to support or oppose broad policies and legislation. Or, as in the case of fetal tissue transplantation research, they may see it set aside altogether in favor of other interests.

Moreover, scientists themselves are not always unalloyed paradigms of purity. "The same human motives that cause other problems in our lives also drive extreme politicized science," writes Michael Gough in *Politicizing Science: The Alchemy of Policymaking.*[4] "There is no surer way to build a powerful bureaucratic empire in a democracy than to promote a supposed peril and then staff up a huge organization to combat it." Gough also notes that the "intoxication of fame and glory" and the realities of greed can distort honest science.

Whatever their frailties, however, most scientists are on guard against any attempt to devaluate their work, or to tamper with, misrepresent or suppress the facts themselves. In *The Republican War on Science*, Chris Mooney defines the abusive politicizing of science as "any attempt to inappropriately undermine, alter, or otherwise interfere with the scientific process, or scientific conclusions, for political or ideological reasons."[5]

The evidence suggests this has happened in recent years. The Union of Concerned Scientists, a nonprofit scientific research and advocacy organization, has been documenting what it calls "political interference in science at federal agencies"[6] and the manipulation of "underlying science to align results with predetermined political decisions"[7] under the second President George Bush.

Since 2004, the organization has also garnered signatures from 12 000 American scientists on its statement, "Restoring Scientific Integrity in Policymaking," which asserts that the administration:

> "has often manipulated the process through which science enters into its decisions… by placing people who are professionally unqualified or

have clear conflicts of interests in official posts and on scientific advisory committees; by disbanding existing advisory committees; by censoring and suppressing reports by the government's own scientists; and by simply not seeking independent scientific advice.... Furthermore, in advocating policies that are not scientifically sound, the administration has sometimes misrepresented scientific knowledge and misled the public about the implications of its policies."[8]

This is not what the American public wants, according to a poll by the Integrity of Science Working Group, a coalition of organizations dedicated to improving the way in which science informs government policymaking.[9] The poll results, released in September, 2004, showed that 84 percent of Americans see an important role in scientific research for the federal government. Of that group, two-thirds believe that government science should be insulated from politics and that experts asked to serve on advisory boards should not be quizzed about their political views.

Roger A. Pielke, Jr., an environmental scientist who studies the use of science in decision-making at the University of Colorado's Center for Science and Technology Policy Research, has tried to put the issue in context. "There is no bright line that separates science from politics," he wrote in a January 2007 statement to the US House of Representatives Committee on Oversight and Government Reform. "In the end what is most important is that the government has the capability to well use expertise in decision-making, because such expertise is absolutely critical to developing, understanding and implementing policy alternatives in the face of the complex challenges of the modern world."[10]

2 From Carter to Clinton: Keeping science at bay

To understand where current approaches to science, policymaking and politics parallel patterns of the past, and where they diverge, it is useful first to consider examples from the administrations of the four presidents who preceded George W. Bush.

The examples presented here illustrate several models of misusing science, but they are neither comprehensive nor an attempt to summarize the overall approach to science under each president. A rigorous analysis would likely show that many administrations used some combination of tactics to deal with "uncomfortable" science.

3 Smothering science: Carter and *in-vitro* fertilization

With the 1978 birth in England of Louise Brown, the first baby to be conceived outside the womb, *in-vitro* fertilization (IVF) and the scientific and ethical debate that surrounds it, gained new attention. IVF enables egg cells to be removed from a woman's ovaries and incubated with sperm in a culture fluid so that fertilization can take place. A few days later, the growing embryo is implanted in the women's uterus, where it can hopefully be sustained for the normal course of pregnancy.

Several years prior to the birth of baby Louise, with Jimmy Carter in the White House, the Department of Health, Education and Welfare (HEW, the predecessor to the Department of Health and Human Services (HHS)) had promulgated regulations requiring that proposals for human IVF research be reviewed by a national Ethics Advisory Board (EAB) before they could receive federal funding.[11,12] HEW moved slowly to establish that advisory board, which meant that most IVF research in the United States was effectively curtailed.

Only in September, 1978, at the instruction of HEW Secretary Joseph Califano, was a newly chartered EAB charged with considering the social, medical, ethical and legal implications of IVF research. Following a year of hearings, expert testimony and input from an array of interested parties, the EAB concluded in its report to the Secretary that "research involving human IVF and embryo transfer is ethically acceptable" and that it might "substantially increase our knowledge concerning the possible risks of abnormal offspring." The EAB also spelled out the conditions on which it could move forward.[13]

Thirteen thousand public comments were received in response, the majority critical of the board's willingness to use federal funds for IVF research. Secretary Califano's office also received 50 letters, signed by 70 members of Congress, opposing IVF research.[11] Much of the opposition centered on the rights of the embryos and the future implications of IVF.

Before the EAB had approved a single federally funded IVF research protocol, a requirement of existing regulations, the Carter administration cut its funding and allowed its charter to elapse. This tactic effectively continued the moratorium on IVF research until 1993, when the requirement for EAB approval was finally lifted.

In effect, President Carter found a quiet way to smother science. The EAB had been permitted to render and publish its findings about the ethics and science of IVF without political interference, and no attempt had been made to dispute them. The treatment itself remained available to infertile women, and the private sector was free to pursue further research at its own expense. But as the Institute of Medicine wrote, "ethically sensitive research is being performed without federal oversight and public input"[14] and the opportunity to conduct publicly funded IVF research was lost.

4 Burying science: Reagan and the AIDS epidemic

In the face of the unfolding HIV/AIDS epidemic, the Reagan Administration impeded science in another way – by trying to bury the problem.

The CDC first reported an unusual infection among previously healthy gay men in June 1981[15] and over the next year the magnitude of the crisis gradually became apparent. The CDC and the NIH formed task forces and sponsored conferences in response to the emerging syndrome, and Representative Henry Waxman held the first congressional hearing, where CDC officials estimated that thousands of people were likely to be affected.[16]

President Ronald Reagan, however, did not say the word AIDS in public until 1985, and then only in response to a reporter's question at a press conference.[17] (That first mention has itself become politicized, with a number of sources claiming that Reagan was entirely silent on AIDS until 1987, the year he made his first substantive speech on the topic.[18])

Other federal officials were told not to speak out at all. Surgeon General C. Everett Koop wrote in his memoir that he was explicitly barred by the Reagan Administration from saying anything about AIDS until 1985. "Whenever I spoke on a health issue at a press conference or on a network morning TV show, the government public affairs people told the media in advance that I would not answer questions on AIDS, and I was not to be asked any questions on the subject," he wrote.[19] In the summer of 2007, Koop reiterated that point at congressional hearings focused on strengthening the office of the Surgeon General.[20]

In *AIDS in the Mind of America*, author Dennis Altman attributes the politicization of AIDS to the fact that already stigmatized populations were the first to be infected with HIV and that the syndrome was firmly linked with sexual behavior.[21] The epidemic also occurred under the administration of a president ideologically committed to cutting federal spending and philosophically partial to the notion of individual responsibility. As Stanley Matek, the former president of the American Public Health Association, said of the Reagan administration, "They tend to see health in the same way that John Calvin saw wealth: it's your own responsibility and you should damn well take care of yourself."[22]

That helps explain why the AIDS research budgets submitted to Congress by the Reagan administration were far less than what the CDC, the NIH and other federal entities of the Public Health Service were seeking. Congress repeatedly intervened with additional resources. In August 1983, the Reagan administration asked for $ 17.6 million and Congress allocated $ 39.8 million. The following year, Congress added $ 30 million to the White House request for AIDS funding, bringing the total to $ 84 million.[23]

Once HIV was identified as the cause of AIDS, public education about transmission routes became the most important tool for reducing its spread. Here, too, the Reagan administration failed to put science to work in the service of health. "The present federal effort is woefully inadequate in terms of both the amount of educational material made available and its clear communication of intended messages," wrote the authors of an Institute of Medicine report. "If government agencies continue to be unable or unwilling to use direct, explicit terms in the detailed content of educational programs, contractual arrangements should be established with private organizations that are not subject to the same inhibitions."[24]

5 Confusing science: Bush I and fetal tissue transplantation

The first Bush administration's approach to fetal tissue transplantation research demonstrated a willingness to confuse, or even undermine, science with unsubstantiated claims.

When Congress sought in 1992 to overturn a 5-year moratorium on federal funding for fetal tissue transplantation research, HHS officials argued that doing so might encourage women to seek abortions. "The reason for the moratorium was our concern not to provide a woman with an incentive to have an abortion," said James Mason, who headed the Public Health Service. "A woman who is ambivalent about abortion might be more likely to choose having an abortion if she believed that the fetal matter might be used directly to benefit the health of another human being."[25]

But the Human Fetal Tissue Transplantation Research Panel had already examined this issue and said in its report that simply wasn't so: "The reasons for terminating a pregnancy are complex, varied and deeply personal... [it is] highly unlikely that a woman would be encouraged to make this decision because of the knowledge that the fetal remains might be used in research."[26] Moreover, the panel noted that 30 years of published studies showed no evidence that the opportunity to donate fetal tissue to research had increased the number of abortions. Nonetheless, the ban on federal funding remained until 1993, a few weeks into the Clinton presidency.

6 Ignoring science: Clinton and clean needle exchange

Like its predecessors, the Clinton administration did not hesitate to make policy decisions that ran counter to scientific findings. But when it came to needle exchange programs for intravenous drug users, it did not dispute the facts.

Under a federal law passed in 1997, Congress required the Secretary of HHS to determine whether making clean needless accessible reduced HIV transmission without encouraging the use of illegal drugs. The law made the availability of federal funds to support such programs contingent on that determination.[27] In a report back to Congress, HHS Secretary Donna Shalala cited reviews of the scientific literature by the US General Accounting Office, the University of California at San Francisco, and the Institute of Medicine, all of which had concluded that needle exchange programs can be "an effective component of a comprehensive community-based HIV prevention effort."[28] Shalala also reported on research showing that needle exchange programs did not encourage illegal drug use.[28] Among the work cited was the NIH's consensus statement, "Interventions to Prevent HIV Risk Behaviors Research," which indicated that such programs may actually reduce drug use by creating opportunities to reach users with treatment and counseling.[29]

Although the Department publicized these findings and encouraged local communities to consider offering their own needle exchange programs, President Clinton declined to lift restrictions on federal funding. The press reported that he had been swayed by General Barry McCaffrey, director of the White House Office of National Drug Control Policy, who had lobbied congressional Republicans to oppose the move and argued that lifting the ban would send the wrong message to children.[30] The President's Advisory Council on HIV and AIDS condemned the decision, with Chairman R. Scott Hitt declaring, "at best this is hypocrisy… at worst, it's a lie. And no matter what, it's immoral."[31] AIDS activists were enraged. Nonetheless, it was a frank policymaking decision that acknowledged scientific realities at the same time it set them aside.

7 Assaulting science: George W. Bush's sharp turn

More than any of its four predecessors, the administration of George W. Bush has been sharply criticized for using every tactic described here, and others, to favor a political agenda over a scientific one. Critics have asserted that all levels of the scientific process – from data gathering to analyzing and disseminating findings to laying a foundation for decision-making – have been under assault.[32–35]

Government scientists have attested to this in surveys conducted by the Union of Concerned Scientists. For example, nearly half (46 percent) of the 279 climate scientists at seven federal agencies who responded to surveys perceived, or had personally experienced, pressure to eliminate words such as "climate change" and "global warming" from their communications. More than one-third (37 percent) said officials at their agencies had made statements that misrepresented scientific findings.[36]

Similarly, almost one-fifth (18.4 percent) of 1 000 FDA scientists who responded to another survey said they had been asked to "inappropriately exclude or alter technical information or their conclusions in an FDA scientific document" for non-scientific reasons. And 61 percent knew of instances where "HHS or FDA political appointees have inappropriately injected themselves into FDA determinations or actions."[37]

As well, the Government Accountability Office found that 6 percent of researchers at NASA, the National Oceanic and Atmospheric Administration, and the National Institute of Standards and Technology had been denied approval to disseminate their research results in the past 5 years.[38] And after an internal review, the Interior Department overturned eight decisions that weakened wildlife protection, decisions made by a political appointee against the advice of department biologists and habitat specialists. The department concluded that the official had improperly favored industry in a manner that violated federal rules.[39]

The process of appointing expert advisors to the many science and technology panels that guide government decision-making also illustrates the Bush administration's willingness to go beyond longstanding practices. To some extent, this is a matter of degree because advisory panels will never be purely scientific entities. "Considerations of politics are unavoidable in the empaneling process," argues Roger A. Pielke, who has called for transparency in the process of selecting scientific advisors, but not for pretending to turn a blind eye to the relevance of political leanings.[10]

But applying a litmus test, as the Bush administration has done, takes the influence of politics to another level. After being nominated to serve on the National Advisory Committee on Drug Abuse, William Miller, a psychologist at the University of New Mexico, reported that a White House official called asking for his views on faith-based initiatives, capital punishment, needle exchange programs, and the legalization of drugs. Miller was also asked whom he had voted for in the last presidential election. Ultimately, he was turned down for the appointment.[40]

The Union of Concerned Scientists has documented many other incidents in which nominees were questioned about their personal beliefs.[41] That line of inquiry, however, defies the recommendations of the National Academy of Sciences, which has said it is inappropriate to ask nominees "to provide non-relevant information, such as voting record, political-party affiliation, or position on particular policies."[42]

In addition to discouraging potentially unfriendly outsiders from participating in the scientific advisory process, the administration has made it difficult for government scientists to express their own views. For example, the HHS now requires scientists to seek permission before they participate on scientific panels of the World Health Organization. An official "notification of foreign travel" must be made 30 days prior to meetings at the World Bank, the Pan American Health Organization, UNICEF, and other entities with a global mission, even at their US-based offices.[43,44]

The Bush administration has pursued other avenues for controlling information as well. A 2003 proposal by the Office of Management and Budget (OMB) entitled "Peer Review and Information Quality" would have subjected government-issued information to a cumbersome external review process if it was likely to have a substantial impact on regulations, public policies, or private sector decisions. Widespread protests by the scientific community lead OMB to retreat on that plan.[45]

But in January, 2007, President Bush signed an executive order requiring federal agencies to have a political appointee in charge of a regulatory policy office that reviews industry guidance documents. "The White House will thus have a gatekeeper in each agency to analyze the costs and the benefits of new rules and to make sure the agencies carry out the president's priorities," reported *The New York Times*.[46] The newspaper said the Environmental Protection Agency and the Occupational Safety and Health Administration were among the agencies of particular concern to the White House.

On the subject of global warning, the Bush administration has been especially aggressive about suppressing or distorting scientific information, according to numerous sources.[34,47,48] In October 2007, the White House removed detailed information about the health risks associated with global warming from the written Congressional testimony of Julie Gerberding, director of the CDC. Gone were specifics about the respiratory problems, infectious diseases, injuries and mental health issues likely to result, as well as numbers indicating how many people would be affected.[49]

Similarly, the White House deleted extensively documented science about the human contribution to warming from a key report by the Environmental Protection Agency on the state of the environment.[48] It also excised material about the impact of climate change on meteorological events from the strategic plan of the US Climate Change Science Program, which brings together top scientists across federal agencies.[47] And in an ongoing effort to highlight divisions within the scientific community, it employed tactics that David Michaels, an epidemiologist at The George Washington University School of Public Health and Health Services, has called "manufacturing uncertainty."[45] The goal is to sow seeds of doubt in order to avoid hard policy choices.

Other distortions have focused on public health education and the dissemination of information. For example, National Cancer Institute officials pulled a fact sheet stating that there is no association between abortion and breast cancer from its web site in November 2002 and replaced it with one that suggested the jury is still out.[34] In fact, the *New England Journal of Medicine* stated categorically that "induced abortions have no overall effect on the risk of breast cancer."[50] In response to a public outcry, National Cancer Institute (NCI) subsequently brought experts together to review all available data and posted a new fact sheet confirming that finding.[51]

Similarly, in 2002, the CDC replaced a fact sheet that explicitly described the proper use of condoms and their value in preventing HIV with one that emphasizes failure rates and abstinence, and offers no instruction in their use.[34] As of September 2007, that fact sheet remained on the CDC web site.[52]

A moral agenda was also on display when the Food and Drug Administration (FDA) rejected Barr Pharmaceuticals' application to make its emergency contraceptive, known as Plan B, available without a prescription. By a vote of 23 to 4, the FDA's scientific advisory committee had recommended its approval for all females, without age restrictions. The FDA overruled the committee, issuing a "not approvable" letter in May 2004 that cited inadequate data on its use by young teenagers.[53]

Barr resubmitted its application to restrict sales to women over 16, but was advised that 18 was the "proper age." FDA then failed to act on the drug for more than 2 years, despite assurances to Congress that it would make a decision. Finally, in August 2006, armed with no more information than it had when the application was first submitted, the agency permitted Barr to bring Plan B to market as an over-the-counter drug.[54]

Embryonic stem cell research is another example of the administration's willingness both to deny science, and to make policy decisions that run contrary to it. In August 2002, President Bush announced that research would be allowed to move forward on 60 genetically diverse stem cell lines that were already available, but that no additional lines could be created. But the existence of those lines was a "fiction," according to a *New York Times* op-ed piece, and the NIH ultimately concluded that just a handful of usable cell lines were actually available.[55]

Since then, President Bush has twice vetoed legislation, in July 2006 and June 2007, which would have made federal funds available for stem cell research. A number of states have stepped in to fill the gap with sizable resources and some researchers have gone overseas to pursue the science.[56,57] But without the support of the federal government, advances are slow, ethical standards are less consistent and may be poorly enforced, and hope fades for a focused national commitment.

8 Options for scientists

Just as presidential administrations have taken many approaches to the use of science in policymaking, so too have scientists chosen many paths to define their own role in shaping the policy. In *The Honest Broker*, author Roger A. Pielke, Jr. identified four idealized roles that scientists might play – he calls them the pure scientist, the science arbiter, the issue advocate and the honest broker of policy.[58]

The pure scientist, writes Pielke, is one who has no interest in the decision-making process, merely providing information and then stepping away. The science arbiter acts as a resource for decision-makers, ready to answer factual questions as they arise. The issue advocate makes the case for a preferred option while an honest broker of policy alternatives clarifies or expands the decision-making choices. "A characteristic fundamental to both Honest Brokers of Policy Alternatives and Issue Advocates is an explicit engagement of decision alternatives (i.e., choices, policy options, forks in the road, etc.)," writes Pielke. "In contrast, the pure Scientist and Science Arbiter are not concerned with a specific decision, but instead serve as information resources."[59]

To some degree, the decisions that government scientists make about the use of their work within a policymaking environment depends on where they see themselves within that framework. Like all employees, they have many conflicting sets of pressures to consider, including concerns about their own financial security and career paths, a professional commitment to advancing knowledge, and an ethical obligation to truth telling. But along with personal preference and specific circumstance, the choice of how best to navigate an uneasy alliance is influenced significantly by their identity as pure scientists or science arbiters, as issue advocates or honest brokers of policy.

Certainly different scientists have responded differently to each of the incidents described earlier. For example, NIH officials, especially those from the National Institute of Child Health and Human Development, worked "quietly, persistently" to end the moratorium on IVF research, urging their superiors in the HEW to reconvene the EAB.[60] Department officials under Carter ultimately agreed to do so, but the plan was shelved when the presidency changed hands.

In *And the Band Played On*, Randy Shilts documents the many contradictory responses of the scientific community to the limited federal resources made available for AIDS in the early days of the epidemic. The same day the HHS secretary told Congress, "I really don't think there is another dollar that would make a difference," a CDC scientist wrote an urgent memo to his boss noting, "the inadequate funding to date has seriously restricted our work.... Our past and present efforts have been and are far too small and we can't be proud."[61] Shilts also described the awkward position in which Edward Brandt, assistant secretary for health, found himself. Brandt testified before a US House of Representatives subcommittee that additional funds for AIDS were unnecessary just days before telling colleagues within HHS essentially the opposite.[62]

Scientists in leadership positions may face particularly awkward dilemmas. Bernadine Healy chose to shift her position on fetal tissue transplantation research when she switched hats from private citizen to government spokesperson. Harold Varmus, who headed the NIH when Clinton denied federal funding for needle exchange programs, was said to have looked "clearly uncomfortable" when the announcement was made, but he did not protest publicly.[31] Julie Gerberding stated publicly that she was "absolutely happy" [63] with her Congressional testimony about global warming, despite the very substantial cuts made by the George W. Bush White House.

Some of the most outspoken opposition to the politicization of science has come in response to the activities of that administration. In July 2007, former Surgeon General Richard Carmona testified before the US House of Representatives Committee on Oversight and Government Reform that "anything that doesn't fit into the political appointees' ideological, theological or political agenda is ignored, marginalized or simply buried."[20] Carmona said that during his tenure, his speeches on stem cells and contraceptives were censored, even where he was merely explaining the science to the public. The administration also blocked release of Carmona's report on global health after he refused to incorporate material about health care in Iraq and Afghanistan into it.[64]

Carmona's term as Surgeon General expired before he spoke out, but other scientists have made public statements while remaining on the government payroll, or they have leaked information to the media anonymously. For example, Drew Shindell, a NASA scientist, testified before Congress that all media interviews were monitored by a NASA press officer and that the Bush administration had watered down his press release on the warming of Antarctica.[65] His testimony was widely covered in the domestic and international media.

Resigning in protest is another option. Rick Piltz, formerly a senior associate of the US Climate Change Science Program, quit in March 2005 after concluding that "politicization of climate science communication by the current administration was undermining the credibility and integrity" of the program. Piltz released a detailed memo outlining his concerns.[66]

Susan Wood, formerly director of the FDA's Office of Women's Health, also protested the Bush administration's approach to science by resigning after the FDA refused to approve the Plan B emergency contraceptive. "I felt there was no role – not just for me but for the people who have expertise," Wood told *The New Yorker*.[35] "I lose a lot of battles; normally you go out and work to fight another day. But this time I just couldn't look in the mirror and live with myself."

In any of these instances, scientists must decide not only what stance to take – a choice dictated in part by how they perceive their roles within Pielke's framework – but also whether to make that stance public. Scientists may elect to speak on the record, for attribution, or they may talk to a reporter on background. In either case, involving the media is likely to bring at least as much attention to the political context of the science, as to the relevant scientific issue itself.

That, at least, is what FDA scientist Rosemary Johann-Liang discovered. In contrast to Susan Wood, Johann-Liang planned to leave the agency quietly after being reprimanded for recommending that the diabetes drug Avandia include a "black box" safety warning.[67] But the incident came to the attention of the media and despite her original intention, she found herself asked to speak publicly about the manipulation of science.

Like individual scientists, the federal government has a number of tools for protecting the integrity of science. In 2007, the Government Accountability Office undertook a survey in response to congressional concerns about the limits being placed on researchers seeking to share their work. After assessing policies in place at NASA, the National Oceanic and Atmospheric Administration, and the National Institute of Standards and Technology, the

Government Accounting Office (GAO) recommended strategies for clarifying dissemination requirements and concluded: "More and more of the major policy debates of the day hinge on the results of scientific research. Therefore, timely and thorough dissemination of research results within the research community and to the public at large is crucial."[38]

Congress, too, has become involved. The Whistleblower Protection Enhancement Act, which has been passed by the House of Representatives and referred to a Senate subcommittee, would explicitly protect federal employees who report that federal research has been suppressed or distorted for political reasons. The legislation defines such "abuse of authority" as including "political interference with science, such as actions that compromise the validity or accuracy of federally funded research or analysis and the dissemination of false or misleading scientific, medical or technical information."[68] Included in the list of reportable actions is "any attempt to suppress the right of government scientists to publish or announce their findings in peer-reviewed journals or public meetings with their fellow scientists."

Regardless of any administrative and legislative action, many scientists on the federal payroll will continue to confront situations in which policy conflicts with the conclusions of their work. They can be given no uniform guidelines to follow in that situation, only a host of difficult choices.

Arguably, government scientists should be held to the highest standard of accountability since they are paid with taxpayer funds to serve the public interest. But does that mean they should speak out publicly when abuses occur, or work quietly behind the scenes, hoping to curb the worst of them? Are the nation's interests best served when someone resigns in visible protest, and the media covers the story widely? Or when they stay on, awaiting a change in the political climate while working to ensure that government science is good science and that knowledge continues to advance?

Ethicists are best qualified to debate the answers, and clearly no universally applicable guidance is possible. But as science advances, playing an ever more important role in the fabric of our society, the interface with policy and politics will undoubtedly grow in complexity as well.

SECTION

4

Bioethics and Science Reporting: Past Trends, the Current Relationship Between Scientist and Journalist, and the Future of Biomedical Publishing in the Age of the Internet

Robert J. Levine, M.D.
Professor of Medicine and Lecturer in Pharmacology;
Director, Donaghue Initiative in Biomedical and Behavioral Research Ethics,
Director, Law, Policy and Ethics Core,
Center for Interdisciplinary Research on AIDS
Yale University, New Haven, CT, USA

Robert J. Levine is Professor of Medicine and Lecturer in Pharmacology and Director of the Donaghue Initiative in Biomedical and Behavioral Research Ethics at Yale University. He is a Fellow of The Hastings Center and the American College of Physicians; a member of the American Society for Clinical Investigation; past-President of the American Society of Law, Medicine and Ethics and past-Chairman of the Connecticut Humanities Council. In the past he was also Chair of the Institutional Review Board at Yale-New Haven Medical Center (1969–2000), founding Co-Director of Yale University's Interdisciplinary Bioethics Center and Chief of the Section of Clinical Pharmacology at Yale, Associate Editor of Biochemical Pharmacology and Editor of Clinical Research. Dr. Levine was the founding editor of IRB: A Review of Human Subjects Research and has served as consultant to several federal and international agencies involved in the development of policy for the protection of human subjects.

He is the author of numerous publications including the book, *Ethics and Regulation of Clinical Research*. In the last 30 years, most of Dr. Levine's research, teaching, and publications have been in the field of medical ethics with particular concentration on the ethics of research involving human subjects.

Dr. Levine has been awarded the Outstanding Achievement Medal from the Office for Human Research Protection, US Department of Health and Human Services, in 2004 for his role in the development of the Belmont Report; the Lifetime Award for Excellence in Human Research Protection from the Health Improvement Institute in 2004 and the Lifetime Achievement Award for Excellence in Research Ethics from PRIM&R (Public Responsibility in Medicine and Research) in 2005.

CHAPTER

12

On the Relations between Scientists and Journalists: Reflections by an Ethicist

Robert J. Levine

The editors of this volume have presented "...several contemporary and historical vignettes..." with the goal of performing "...a postmortem examination of what went wrong, and what we can learn from past mistakes...."[1] This chapter presents several additional historical vignettes, mostly drawn from my personal experiences; each suggests considerations of the "ethical dance between scientists and the media" in addition to those provided by the editors. I will also speculate on why academics and scientists are so often misunderstood by journalists and others. I believe there are essential features of the academic and other professions that 'non-professionals' do not and, perhaps, cannot appreciate.

Before proceeding with my vignettes, I will comment briefly on this book and the conference at which its chapters were presented. The objectives of this book, as set forth in the editors' introduction, are stated primarily in the form of questions beginning with the words *what* and *how*. "*What* are the causal factors that lead to inaccuracies and misleading interpretations, in the reporting of biomedical discoveries...." and "*How* effective is the biomedical research community at policing itself...." These are empirical questions and, as such, lie outside the bounds of an ethicist's expertise. Ethical questions typically begin with such words as *ought* or *should*. Sometimes *should* is implicit as in "What ethical obligations do scientists have when we convey our research findings...." This, the principal normative objective identified by the editors, could be restated so as to make it more easily identified as such: "What *should* scientists do when we...."

The major normative issues in this field are reasonably well understood and agreed upon. Scientists and other scholars have an ethical obligation to present an accurate account of the fruits of their efforts whether these presentations are to their colleagues or to the larger general public. This responsibility includes making reasonable attempts to explain the context within which their work can be adequately understood. It further includes making reasonable attempts to convey the limitations of their work; this entails considerations of such matters as the relevance of animal models, the probative value of preliminary data, the limitations of subject populations and the like.

What counts as a reasonable attempt? Who is responsible for seeing to it that such attempts are made satisfactorily? These are the subjects of several other chapters in this book.

1 Historical vignettes

1.1 The scientist's urge to publish

- In 1968, I was invited to present a paper at a symposium in Atlantic City at the annual national meeting of FASEB (The Federation of American Societies for Experimental Biology). My presentation consisted of evidence that histamine was the mediator of acid secretion by the stomach. About one month before the meeting I received a letter requesting that I forward 150 copies of my paper for distribution to the press. I declined to do so on grounds that my experiments had been done almost exclusively on rats; my evidence that these results also applied to humans was, at this point, very preliminary. I believed it was premature to issue a press release because the meaning of the data was likely to be misinterpreted.

By return mail I received an invitation to meet personally with the woman who had sent me the invitation; she was in charge of management of the FASEB meeting. When I met with this remarkable woman, herself an accomplished scientist, she explained she wanted to meet me because she had never before had such a request refused. "Usually," she explained, "a request for 150 reprints for publicity purposes yields 300."

- Approximately fifteen years ago, during my term as chairperson of the Institutional Review Board (IRB) at Yale-New Haven Medical Center, I was called by Dr. B, a very distinguished member of our faculty, who asked me about getting IRB review and approval of a research project. The project in question had been carried out by investigators in another country. After its completion, they drafted a manuscript and sent a copy to Dr. B. soliciting his comments. After he provided these comments, they decided to recognize his contribution by adding his name to the list of co-authors. The editor of the journal to which they had submitted the paper, assuming that authorship in this case meant what it usually does, informed the authors that the journal would not review the article until the editorial office received documentation of its having been approved by the IRB at Dr. B's institution, Yale University. I refused on grounds that it was the IRB's mandate to review research before the work was done. (Personally, I thought his efforts merited an acknowledgement, not co-authorship.)

2 Faulty understanding

- In 1992, a woman in Florida gave birth to an anencephalic baby. Her offer to donate its tissues and organs for transplantation purposes was denied by a judge. The next morning I was quoted on page one of *The New York Times* as having said this baby was "...more like a fish than a person." Needless to say, I was extremely distressed. I tried repeatedly and without success to reach *The Times* editorial offices by telephone; I wanted to publish a letter to the editor that would correct my statement.

Late in the morning I received a phone call from another *Times* reporter requesting my comment on another story he had published in the same issue of *The Times*. I said I was so distressed by the fish misquote that I had not read any further in the paper. He asked me to wait 10 minutes for him to "fix things". Shortly thereafter he called with instructions to call another number at *The Times* where the National Desk Editor would be prepared to receive my call. When I called a woman who said she was an assistant editor asked me to tell her what I actually had said. I told her that in the context of my lengthy explanation of anencephaly, I concluded by saying that "the anatomical configuration of the brain of an anencephalic newborn has more in common with that of a fish than it does with that of a normal human newborn." Upon hearing this, the assistant editor asked, "What's the difference" between the two statements?

I asked her to imagine that during the 2nd trimester of her pregnancy she was told that she was carrying an anencephalic fetus and that she rejected a recommendation to abort because she decided instead to donate the tissues and organs for transplantation purposes. Further, having chosen delivery by Caesarian section to avoid damage to the organs, she was then told by a judge that she was forbidden to carry out her plan. And then, the next morning, she reads that an ethicist of whom she has never heard has proclaimed her baby "more like a fish than a person."

"OK," said the assistant editor, "we will print your letter." And they did.

Later that afternoon I was called by another *Times* journalist who said to me "That's a great line – the sort that gets us on page one above the fold. In all the years I've been interviewing you, you have never given me such a quote. You are always so cool and measured, it's all I can do to get you in the classified section."

I replied that her colleague had not gotten that quote from me. "That's what I figured," she replied. "What did you really say?"

- Faulty interpretation of the meaning of newspaper accounts of scientific results is not limited to non-scientists. Many years ago I received an urgent call from a distinguished scientist who was a member of the leadership of a multi-centered clinical trial of a nutraceutical; he had just read a newspaper report that a clinical trial of this agent conducted in Europe yielded an incidental finding that its administration to those who smoked was associated with a 17 percent increase in the annual incidence of lung cancer. "That's awful," he said. "What are my ethical obligations in the face of this information? Should I stop the trial? Remove all cigarette smokers? Renew the informed consent of each subject?"

I asked for an opportunity to examine the data from the trial. They showed that the annual incidence of lung cancer among smokers who received the nutriceutical was approximately 1 in 6,000 and that the incidence among smokers in the control group was about 1 in 7,000. Upon hearing this explanation, my distinguished colleague said that the increase was so small it was scarcely worth worrying about. I replied that although it looked small, it was a 17 per cent increase – just as the journalist reported.

3 Deliberate disinformation

Following are some bits of disinformation presented at meetings of NHRPAC (National Human Research Protection Advisory Committee) by a member of the NHRPAC and reported in the press even though they were corrected at the same meeting by another NHRPAC member[2]:

- *Disinformation*: Fen-Phen® administration leads to the development of clinically significant heart valve lesions. The active ingredient in the causation of these lesions is fenfluramine. A team of scientists has administered fenfluramine to normal healthy children for research purposes. This was an outrageously dangerous and thus unethical thing to do to normal healthy children.
- *Facts*: It is correct that Fen-Phen® administration has been associated with heart valve lesions and that there have been cases of cardiac lesions in which fenfluramine was the only agent of the combination (of two ingredients) that was administered. However, development of such lesions has been associated only with prolonged administration of either fenfluramine or Fen-Phen® at much higher doses than those used for research purposes. The research use in question was only of single small doses which have never been associated with cardiac lesions.

- *Disinformation*: Food and Drug Administration required administration of a dangerous pediatric preparation of an antibiotic to healthy children. This was highly unethical because it exposed the children to grave danger with no possibility of direct benefit to the children.
- *Facts*: The research in question called for administration of the flavored liquid in which the antibiotic was to have been suspended. This was, in essence, a test of "consumer satisfaction" with the flavor of the vehicle. No healthy child received any antibiotic in this study.

4 Mischaracterizations

- The National Institutes of Health (NIH) was accused of withholding sterile needles from Intravenous Drug Users (IVDUs) (persons who use illicit drugs administered intravenously) who wished to participate in needle-exchange programs. This was portrayed as a heinous act in that it exposed these persons unnecessarily to the risk of contracting HIV infection. This charge was leveled at a NIH-sponsored randomized clinical trial comparing the outcomes in IVDUs assigned to free needle exchange with those in IVDUs who were enabled to purchase clean injection equipment from private pharmacies at a nominal price and without a prescription.[3] There were substantial preliminary data indicating that many IVDUs preferred the pharmacy approach to free needle exchange; their reasons included shorter waits and better safeguards of privacy and confidentiality.

These allegations were the subject of extensive coverage in newspapers and on television. Press coverage included the statements of the complainers that the NIH Director was a "murderer".

A committee of bioethicists was appointed by the NIH Director to review the charges and the program.[4] At a meeting of the NIH Director's Advisory Committee this Committee reported its finding that the charges were a deliberate mischaracterization of the facts of the case and that the clinical trial was ethically justified. During the "coffee break" which followed this report journalists interviewed the lead complainer who repeated the same mischaracterizations; these were featured in the following day's newspapers – not the report of the committee.

- At a public meeting, a federal official who was criticizing the American system for protecting the rights and welfare of human research subjects, pointed out that no-one could inform him of how many humans served as research subjects within any year. He contrasted that with the fact that he was able to get precise information on the number of animal subjects. Why, he asked rhetorically, do scientists care more about animals than they do about people?

Speaking as a representative of human subject protection systems, I replied to each of these points:

1 We don't count subjects because no one had ever before suggested that this would contribute to protecting their rights or welfare. I proposed that if he wanted to see highly meticulous records of numbers and other descriptors of human research subjects, he should examine the records kept by the Nazi researchers in the nefarious concentration camps during World War II;
2 We know the exact numbers of animals because we purchase them with money we receive from such sponsors as NIH. They require a detailed account of how we spent their money. Our methods for recruiting human subjects are much different; and
3 We don't care more about animals than we do about people.

A journalistic account of this story published the following day quoted the federal official's charges but none of my responses.

5 Lawsuits

- Next I will mention two class action lawsuits the initiations of which were heralded by spectacular coverage in the national press.[5] In addition to banner headlines in newspapers, the first was the subject of a *60 minutes* television presentation; the author of the other received an award for outstanding investigative reporting. The first, a double-blind, placebo-controlled study of the withdrawal of antipsychotic medication from patients with schizophrenia is commonly referred to as "the UCLA schizophrenia-placebo case".[6] The second alleged, among other things, inadequate informed consent in a study of therapy designed to prevent graft versus host disease in recipients of bone marrow transplantation at the Fred Hutchinson Clinic.[7] In recent years I have asked the members of many audiences of persons concerned with clinical research, its regulation and its review by IRBs to raise their hands if they were aware of these cases; nearly all of them have done so. Then I have asked those who knew of the final outcomes of these cases to raise their hands; virtually none did.

The first case was settled for $199 000, the amount the defendant institution calculated it would cost to win the case if it went to trial. This means that the plaintiffs received nothing whatever. According to estimates I consider reliable, the plaintiffs' attorneys sustained a loss of approximately $2 000 000.00. If the plaintiffs' believed the extravagant claims made at the outset, one would think they would not have agreed to such a settlement.[8]

The second case went to trial. The jury awarded $1 000 000.00 to one of the plaintiffs – the person whose bone marrow had been lost when a tube containing it broke in a centrifuge; this is the amount the Clinic had offered as a settlement to this individual before the case went to trial. There was no award to the others.

As is typical of such cases, their ultimate resolutions received little or no media coverage.

6 Differing perspectives

- When the media reported that Vioxx® administration was associated with an increased incidence of heart attacks and, further, that the industrial sponsor of this drug had long been aware of this and had withheld this information from the public and from practicing physicians, most readers felt anger: "Here we go again, yet another case in which the rapacious drug industry lied to the public in order to maximize profits." Apparently, the excess incidence of heart attacks had been observed by the data and safety monitoring committee (DSMC) and they had delayed for quite some time reporting this information to the sponsor. People who had experience in the field of drug development, particularly those who had served as members of DSMCs, had entirely different initial reactions. They sympathized with the DSMC members. They knew how difficult it is to interpret the meaning of early trends in adverse event data during a randomized clinical trial. It is particularly difficult when, as in this case, preclinical and phase one data do not anticipate the particular adverse event and when the adverse event is a common occurrence in the population serving as research subjects. Moreover, in this case some of the control subjects received as the comparator naproxen, a drug that was at the time suspected of having an aspirin-like protective effect against heart attacks.[9]

7 Speculation on the cause of misunderstanding of scientists and other professionals

Why do scientists, scholars and physicians have such difficulty making their work understood by journalists and others who are not members of their professions? It is often said that they could be better understood if only they would make more of an effort. In the field of informed consent, for example, they have been encouraged to present information at a sixth grade reading level.[10] I think that the barriers to understanding are much more complex and not susceptible to easy resolution through simple translation.

I think that we can begin to understand the nature of the problem by reflection on Alasdair MacIntyre's concept of 'practice':[11]

MacIntyre defines a practice as:

> "...any coherent and complex form of socially established cooperative human activity through which goods internal to that form of activity are realized in the course of trying to achieve those standards of excellence which are appropriate to, and partially definitive of, that form of activity...."

MacIntyre distinguishes the 'goods' accomplished by those who engage in the practice as either external or internal. Among the external goods are prestige, status and money. There are always alternative ways for achieving such goods; "...their achievement is never to be had only by engaging in some particular practice." By contrast, "...there are the goods internal to the practice...which cannot be had in any way..." other than engaging in the practice or some other relevantly similar practice. These internal goods "...can only be identified and recognized by the experience of participating in the practice in question. Those who lack the relevant experience are incompetent thereby as judges of internal goods." Moreover, only those who have submitted to the tutelage and apprenticeship of accomplished practitioners can understand the nature of internal goods.[12]

To illustrate, I will offer a brief explanation of an internal good in the practice of medicine. In medical practice, one must begin by submitting to the tutelage of accomplished practitioners; one enrolls in medical school and then continues the apprenticeship during internship, residency and postdoctoral fellowships. In this apprenticeship one acquires through multiple repetitions under the guidance of experienced practitioners the ability to use a stethoscope to hear and interpret the meanings of various sounds made by the heart. This ability may be considered a *virtue*, "...an acquired human quality the possession and exercise of which tends to enable us to achieve those goals which are internal to practices and the lack of which effectively prevents us from achieving any such goods."[13] Armed with this and other relevant virtues, the competent medical practitioner arrives at a diagnosis and a suitable plan of action. The process culminating in a diagnosis and a plan of action that fellow practitioners will approve is an internal good. It can be fully appreciated and understood only by fellow practitioners. The idea of excellence that can be fully appreciated only by fellow practitioners is embodied in the physicians' characterization of the pinnacle of professional accomplishment – to be selected by fellow practitioners to serve as their physician. The ultimate expression of admiration is to refer to a fellow practitioner as a "doctors' doctor."

Those who have not been acculturated into the practice of medicine necessarily have great – often insurmountable – difficulties in appreciating just what it is that causes physicians to single out some of their colleagues as "doctors' doctors". By contrast, it is easy for the non-member of the practice to recognize the external goods of medical practice – the physician's income, the source of such income, the hospital's survival rate for open heart surgery, etc. This, I believe, is why current commentary on the ethics of medical practice is so focused on the external goods of the practice – conflicts of interest, etc. The internal goods – the true indicia of excellence – are, perhaps intractably, beyond the comprehension of the general public.

The practices of the various sciences are similarly composed of internal and external goods. It is, perhaps, more difficult to give brief descriptions of the virtues and internal goods of the sciences than it is for the practice of medicine.[14] Just as with medicine, and, I believe, for similar reasons, public commentary on the sciences tends to focus on the external goods of its practices – those easily identified by non-scientists, including, but not limited to conflicts of interest, plagiarism and some, but not all, types of data-faking.

8 Summary

This chapter begins with brief presentations of stories of interactions of scientists and physicians with journalists. These stories, mostly drawn from my personal experience, were selected to illustrate ethical issues in addition to those highlighted by the editors in their introduction to this book. The issues include the perhaps excessive desire of some scientists for publicity, unintended faulty understandings of information, deliberate dissemination of disinformation and mischaracterizations, imbalanced coverage of certain news events (e.g., lawsuits) and differing interpretations of 'meanings' according to such things as the reader's professionalization.

Why do scientists, scholars and physicians have such difficulty making their work understood by journalists and others who are not members of their professions? In this chapter, I present my speculation on the cause and nature of this lack of understanding. Drawing on Alisdair MacIntyre's concept of *practice,* I suggest that extensive misunderstanding is inevitable and, perhaps, intractable owing to the inability of most persons who are not members of a practice to appreciate the *internal goods* of that practice. Those who do not appreciate the internal goods of a practice are necessarily limited in their ability to report on the activities of its practitioners.

Ms. Laura Spinney
Freelance science writer
London, UK

Laura Spinney is a freelance science journalist and novelist based in London and Paris. Her training consists of a B.Sc. in Natural Sciences from Durham University, a postgraduate diploma in periodical journalism from the London College of Printing and Distributive Trades, stints at *Vogue* and *The Catholic Herald* in London, and *The East African* newspaper in Nairobi, and finally an internship at *New Scientist* (London) in 1995. Since 1995 she has contributed to the science pages of *The Guardian*, *The Independent*, and *The Daily Telegraph*, of the British broadsheet newspapers, and she now writes regularly for *New Scientist*, *The Economist*, and *Nature*, among many other publications. Her first novel, *The Doctor*, was published by Methuen in the United Kingdom in 2001, and her second, *The Quick*, will be published by Fourth Estate in March 2007.

CHAPTER

13

Don't Shoot the Messenger

Laura Spinney

In 1994, the popular science magazine *New Scientist* described the work of a chemist at Nottingham University in the United Kingdom, Martyn Poliakoff, who was experimenting with supercritical fluids as a novel type of solvent. These fluids flow like gases, dissolve substances like liquids and have none of the environmental contamination problems of traditional solvents. According to the article's author, David Bradley, Poliakoff received a call soon after it was printed, from a British chemical manufacturer called Thomas Swan & Co Ltd. That call marked the beginning of a collaboration which resulted, in 2002, in the construction of a radical new type of chemical reactor. Not much bigger than a cornflake packet, as efficient as a reactor 250 times its size, it was the epitome of green chemistry.

In March 2006, *The Guardian* published an article by the Russian journalist Anna Politkovskaya, who was to be so tragically killed in Moscow just a few months later. The article strongly suggested that an outbreak of mysterious illness in schools in Chechnya was the result of mass poisoning, not mass hysteria as the Russian authorities had claimed it was. Politkovskaya was an outstanding journalist, but on this occasion her instincts let her down. No toxin was ever found, and a program of psychological rehabilitation was what eventually cured the children, some of whom spent more than 6 months in hospital. According to Khapta Akhmedova, the Chechen psychologist who treated them, they were indeed suffering from mass hysteria.

I give these two examples to demonstrate the power of the press. The first is positive: it shows how good journalism can bring people together who might not otherwise meet, and stimulate collaborations which result in concrete benefits for humanity. The second is negative, for reasons that are less obvious. Mass hysteria has been documented since the Middle Ages. In 1987, the British psychiatrist Simon Wessely suggested that it came in two forms: acute episodes that vanish quickly once their true cause has been acknowledged, and a more serious, chronic variety that tends to take place against a backdrop of social trauma, such as existed in Chechnya at the time Politkovskaya was writing (and that still exists there today). "For an episode to become chronic," says Wessely, "It has to be believable by those affected, and it has to be reinforced, at least at the start, by local experts, including physicians and the media." Without intending to, Politkovskaya may have hurt the very people whose plight she sought to bring to the world's attention.

To state some self-evident truths and then, hopefully, some less self-evident ones, science journalists have a different agenda from scientists, governments, lobby groups and industry, and that is to present the facts as fully and objectively as possible. They exist because there is a demand for them, because much research is funded by taxpayers; that research may affect those taxpayers' lives, and they therefore have a desire and a right to know about it. Those taxpayers include, of course, scientists, lobbyists and government ministers. And just as it is those people who create the demand for journalists, it is they who, at least to a certain extent, dictate how many journalists there are, how many column inches they fill and what they fill them with. I'm talking about print journalists, but the rule holds across all media, and only slightly less so for public service broadcasters such as the British Broadcasting Corporation (BBC) than for commercial outfits.

These are three of the basic rules one learns at journalism school: "man bites dog" is a story, "dog bites man" is not. A news story should have the structure of an inverted pyramid, with all the essential information in the first paragraph and the information content diminishing with each subsequent paragraph – this so that an editor can cut from the bottom without harming the story's integrity (contrast scientific papers, which save their conclusions for last). And tabloid journalists work for the largest share of the readership and are therefore the most laudable members of our profession. In the immortal phrase, a journalist is only as good as his or her last story. Our incomes and reputations depend on our ability to feed our editors a constant stream of spellbinding copy. We must inform, but we must also entertain. The news machine is never satiated and it prefers controversy. Maybe it's a relic of some ancient survival mechanism that has to do with detecting a threat before it detects you, but people are more likely to read a story with the headline, "Clinical trial of new drug goes spectacularly wrong," than one entitled, "Clinical trial of new drug passes off smoothly." They just are.

Given that science journalists work for their readers, there is bound to be a tension between them and the scientists whose work they write about. This tension is often depicted as unhelpful. In the United Kingdom, a House of Commons science and technology select committee concluded in 2006 that "the media is [sic] seen as part of the problem rather than the solution to improving public understanding of risk and scientific advice".[1] Journalists get it wrong sometimes, occasionally with grave consequences as I will discuss later. I will even suggest that there are some fundamental aspects of the way they work that need to be changed. However, this perception must be an example of how all human beings, not just journalists, prefer to linger over bad news.

"Professor Sinsheimer is confident that the structural nature of the gene – at present a rather fuzzy, abstract idea – will involve the interplay of protein and DNA." That sentence first appeared in *New Scientist* in July 1957, referring to comments made by a biophysicist in Iowa named Robert L. Sinsheimer. How quaint it seems now, when most well-informed people, even if they have had no scientific education, know that a gene is a stretch of DNA in the form of two intertwined helices, that this DNA is translated to generate a protein, and that it is through genes that familial traits are inherited. Maybe that knowledge will even be useful to them one day – if they find themselves in need of genetic counseling, for example – and who is responsible for them having that knowledge, apart from teachers, if not science journalists?

I don't think it is unreasonable to suggest, either, that science journalists had a hand in persuading the South African government to alter its position on HIV treatment, away from advocating a diet of garlic and beetroot and towards the use of anti-retroviral drugs – the only treatment for HIV infection that is currently recognized by the mainstream scientific community. Or that science journalists helped bring the debate over human-induced climate change to the public, so forcing national leaders to confront it as a potential problem. Perhaps that confrontation was slow in coming, and I will explore some of the reasons for that later on, nevertheless it has now happened, and leaders are altering their policies with respect to the environment because of it.

1 Why scientists need the media

Having declared my interests as a journalist, I'd like to suggest that the tension between the media and scientists can be fruitful. First, and most obviously, by persuading readers that what scientists do is worth supporting, both morally and financially. The Bush administration's stranglehold on federally funded stem cell research shows what can happen when voters (readers, viewers, listeners) decide it isn't. Second, because scientists tend only to be experts in a narrow field, and a molecular biologist may not know much about particle physics, the media can inform them about developments in those distant fields and so stimulate interdisciplinary cross-fertilization. That cross-fertilization needn't be limited to science, either. The collaboration between Martyn Poliakoff and Thomas Swan, which resulted in a new, environmentally friendly chemical reactor, is an example of a highly successful partnership between academia and industry that was inspired by the popular press.

Third, scientists should not be the ones to decide how society should apply their discoveries, except as members of that society. Most of them sensibly shun this responsibility anyway, and then there are the megalomaniacs. The media provide a forum for discussion about the applications that are acceptable to a given society, and those that are not. For example, as I reported in *The Guardian* on March 17, 2005, free range hens may be seen as the ethical alternative to battery hens, but they have problems of their own – problems that should make us think twice about the way they are reared.

For reasons that are not clear, free range chickens have a tendency to peck at each others' feathers and in extreme cases, to cannibalize. In 1985, Kim Cheng of the University of British Columbia observed that congenitally blind hens did not do this, and that blind hens might even be less susceptible to stress than seeing ones. He suggested breeding blind, free range hens as a solution. There is no scientific evidence that blind hens are worse off than seeing ones in animal welfare terms, but Peter Sandøe, a bioethicist at the Royal Veterinary and Agricultural University in Copenhagen, Denmark thinks that most people would still consider it wrong to breed them. "In general, science gives important but often rather narrow insights into animal welfare," he says. Sometimes science doesn't offer any insights at all, and an arbitrary decision has to be taken. In 2000, for example, a European Commission report recommended that 30 kilograms per square meter was the maximum density at which broiler hens should be kept. But, says Sandøe, "There is clear evidence suggesting a linear relationship between density and stress [in broiler hens], so there is no obvious cut-off point."

A similar sliding scale argument applies to states of consciousness. Persistent vegetative state (PVS) is diagnosed when patients have remained in a coma for a protracted length of time, and since PVS patients are regarded as unlikely to recover, they can be considered eligible for euthanasia. In 2002, however, a new state was identified, called minimally conscious state (MCS), that slots between severe disability and PVS on the consciousness scale. MCS patients are able to respond to an external stimulus, for example they might smile when a friend enters the room, and the medical community currently considers them potentially capable of recovery – and therefore not eligible for euthanasia or organ harvesting. However, MCS is as difficult to diagnose as PVS, and the case of the brain-damaged American woman Terry Schiavo, who was at the centre of a right-to-die battle in 2004, illustrates the confusion that reigns over these definitions. Sometimes the science creates more questions than answers, but since decisions like the one to withdraw Schiavo's feeding tube will have to be made more and more often, we need to have the public discussions that will help to create the ethical framework into which new findings will fit, as and when they come along.

Fourth, the press can highlight the plight of scientists when attempts are made to censor them. In July 1976, *Le Monde* described how Chile had lost 18 percent of its doctors and 30 percent of its engineers in an "exodus of brains" that had begun 6 years earlier. The majority of the engineers left under the Allende regime, while his successor General Pinochet removed most of the basic scientists and sociologists from their posts. In 1999, a UK-based publication called *Index on Censorship* warned that, as science became increasingly corporate (including government-funded), censorship would become less about firing scientists or locking them up, than about being economical with the truth. In 2006 the same publication reported that US government officials may have tried to stop government scientist Thomas Knutson from speaking to the media about his views on global warming. Censorship is alive and well; and an inquisitive, skeptical media are one of the best weapons we have against it.

Fifth and perhaps most controversially, I believe that at its best, scientific journalism can provide a useful complement to peer review. I am not suggesting that it should set itself up as a rival arbiter of scientific quality. The distinction between published and unpublished work is an important one and unpublished data should be handled with care, as the Royal Society (RS), the UK's national academy of science, emphasized in its 2001 guidelines for the communication of science.[2] However, I believe there is also room for unproven ideas, even wildly fanciful ones, to be discussed openly in a public forum, for the following reasons.

Many scientists acknowledge that peer review, while being the best system currently available for judging scientific quality, is far from perfect. An international conference organized by the European Science Foundation, the Czech Science Foundation and the European Heads of Research Councils in October 2006 concluded that it stifled novelty. "The current system of peer review can at times be considered tyrannical in its approach and it needs to evolve," said John O'Reilly, the chief executive of the UK's Engineering and Physical Sciences Research Council.

Peer review is the vehicle of science by consensus, but the leaps in imagination that bring about revolution and paradigm shift are often taken by mavericks – mavericks like Barry Marshall, who swallowed *Helicobacter pylori* bacteria to prove that they caused stomach ulcers, and in 2005 won the Nobel prize in physiology or medicine with his colleague, Robin Warren. It is striking that in its summary checklist for print and broadcast journalists, the RS encourages us to ask: "Do the researchers have an established track record in the field and are they based at a reputable institution or organization?" Woe betide any struggling geniuses in, say, the resource-poor European Union accession countries. Marshall was a junior doctor when he first took on the field of gastroenterology. Josef Loschmidt was a secondary school teacher in Vienna when, in 1861, he published the cyclic structure of benzene. Completely unknown as a scientist, unable to get his work published in a recognized journal, he paid a Viennese publishing house to produce a small print run of his pamphlet. Four years later the renowned chemist August Kekulé published the same cyclic structure – and his is the name we remember.

Every year, the British Association for the Advancement of Science (BAAS) organizes a science festival for the public. According to Ted Nield, editor of *Geoscientist* magazine and current chairman of the Association of British Science Writers (ABSW), at the 1923 BAAS meeting British geologists discussed the theory of continental drift, which in the US at the time was universally rejected on the grounds that no known physics could make it work. The first direct measurements of continental drift were not released by the American space agency NASA (National Aeronautics and Space Administration) for another 60 years, but as Nield wrote in the ABSW newsletter in October 2006, "At no time did the opposition of the scientific establishment ever hinder the BAAS discussing these ideas."

If scientists feel confident enough to discuss ideas for which there is as yet little or no evidence, in front of a lay audience, why should science journalists not report on those discussions, as long as they emphasize that lack of proof? Or to look at it another way, if those public discussions are taking place and journalists don't report on them, aren't they practising a sort of self-censorship? It should be possible to present novel ideas in an open-minded and suitably skeptical way, without being accused of acting irresponsibly. In doing so, journalists might even be instrumental in engaging people with science at a deeper level than they currently are, by demonstrating the creative process at its core and the flashes of inspiration and intuition that drive it in, so to speak, real time.

Science journalists are themselves divided over whether they should cover this kind of debate. Nield made his comments as a preamble to a discussion of an interesting incident that occurred at the 2006 BAAS festival. The association organized a press conference at which a panel of scientists presented their research on telepathy and out-of-body experiences. The science correspondents of many of the British papers were there, and some of them later drummed up comment to pour scorn on the findings – comment which came from academics who were not present at the conference and so had not taken part in the discussion. Some in the media saw this as a laudable case of skeptical journalism, others as a disgraceful example of journalists setting themselves up as arbiters of good scientific taste, which is not their job. Either way, their condemnation meant the telepathy researchers received a great deal of publicity, because those journalists who had decided the research warranted no coverage at all were no longer able to ignore it.

2 When the media get it wrong

It's a question of balance, and sometimes journalists get the balance wrong. Take the example of chronic fatigue syndrome (CFS). Because so little is known about what causes this condition, there have been long-running and sometimes vituperative disputes between clinicians, and between clinicians and patient groups, over whether it is a neurological or a psychiatric disorder, whether it should even be diagnosed as an illness and how it should be treated. When *New Scientist* reported the death in 2006 of 32-year-old Sophia Mirza, as the first time that CFS had been recorded as a cause of death in the United Kingdom, it described the coroner's verdict as a "breakthrough" for those who argue that CFS is a physical rather than a psychological condition. It included no skeptical comment and none of the scientific background to the story, except for one sentence describing the causes of CFS as controversial. The story was picked up by the BBC, and there is now even a Wikipedia entry on Mirza – both (at the time of writing) equally unquestioning. One psychiatrist I spoke to felt strongly that this was a derogation of duty on the part of the magazine, with potentially grave consequences for those working in an already difficult field, and I am inclined to agree.

The reporting of the measles, mumps and rubella (MMR) scandal was a more spectacular flop. On 28 February 1998 *The Lancet* published a paper by Andrew Wakefield, a doctor at the Royal Free Hospital in London, and colleagues, suggesting that there was an association between autism, bowel disease and the combined vaccine against MMR. That paper made no claims of a causal link between autism and MMR, but Wakefield later told reporters that he believed children should receive the three components of the vaccine separately. No empirical evidence has since been found to support a causal link, but his statement fuelled public anxiety and vaccination rates dropped dramatically across the UK. Even now, 8 years after the paper's publication, many people are unsure if it is safe to vaccinate their children with the MMR vaccine.

Whether or not *The Lancet* should have published Wakefield's paper is not an issue I will discuss here, but Wakefield's statement to the press was probably unwise. At the time, he must have looked to journalists very much like one of those mavericks whose virtues I was extolling earlier, and this is of course the dilemma – maverick geniuses, crackpots and the simply misguided all look and sound very much alike. What was unforgiveable about the press coverage was that the evidence became buried under a mountain of comment – some knowledgeable, some not so knowledgeable – so that readers perceived a debate with two sides. In a 2003 survey, the UK's Economic and Social Research Council (ESRC) found that the single most, well-recalled fact from the MMR episode was Prime Minister Tony Blair's refusal to say whether his son Leo had been vaccinated.[3]

Why did this happen, and can we learn any lessons from it? In its 2001 guidelines, the RS stated that in the United Kingdom, specialist scientific journalists are often sidelined by non-specialist editors when it comes to the coverage of controversial issues.[2] The ESRC report found that at the height of the MMR crisis, only 20 percent of the stories were written by specialists, while a UK House of Lords science and technology select committee found that during the genetically modified (GM) "Frankenstein foods" scare, which broke in February 1999, "science correspondents never contributed more than 15 percent of the total news coverage at any stage during the period".[4]

Ben Goldacre, a doctor who writes a science column in *The Guardian*, summed up what happened in the MMR case as follows: "Parents found themselves in the strange position of receiving advice on complex issues of immunology and epidemiology from lifestyle columnists."

There are historical reasons for this, notably that many of the big British papers have leader-writers, mainly from humanities backgrounds, who cherry-pick the most controversial news stories and then express an opinion on them – which is after all what they are paid to do. Unfortunately, however, few of them are equipped to interpret complex scientific evidence. Among general news publications, *The Economist* is a model of good practice in this regard. "Leaders on scientific subjects are normally written by science journalists, and those on other subjects that have some scientific component can be, and usually are, scrutinized by the science editor," says *The Economist*'s science editor, Geoff Carr. "My own experience is that my colleagues are very keen to get the science correct, and actively solicit this scrutiny."

I'm told that this, more rational practice is also widespread in the United States, which may be one reason why the furore over MMR was more muted there. However, even in the United States, a sizeable minority of the population still believes that global warming remains an unproven theory, and though that is partly due to the muzzling of government scientists and the spread of disinformation by industry-funded groups, journalists' inability to make sense of and present the evidence effectively almost certainly contributed.

3 Hall of mirrors

One of the most obvious ways in which readers are let down by a flawed reporting system is through the inadequate communication of risk. The RS has given as an example the tendency of some journalists to focus on a "small but previously unrecognized side effect of a preventative medicine," which is given greater coverage "than the already known greater threat to health posed by the disease the medicine is intended to prevent." It very sensibly advises journalists to give a measure of absolute risk as well as one of relative risk – that is, to spell out that a 30 percent increased risk of stroke for those taking a certain drug (which sounds scary, and is a relative risk) corresponds to one extra stroke in every 5 000 people taking the drug (the absolute risk, which sounds less scary).

Goldacre gave a nice example of the relevance of maths to political or any other kind of decision-making in his *Bad Science* column on 9 December 2006. He was responding to a newly published British government report that claimed that one murder a week was committed by a psychiatrically ill person. This prompted kneejerk calls from some quarters for preventive detention of dangerous psychiatric patients. By walking us through the maths in a way that entertained without ever condescending, he demonstrated the difficulties in predicting an extremely rare event such as murder, even when a highly predictive test is available (it isn't). This is because the low background frequency of the event means that the number of false positives generated by the test becomes non-negligible compared to the number of genuine positive results, causing the test's predictive value to fall away. Without explicitly drawing any conclusion from his explanation, he highlighted the absurdity of a policy of preventive detention in these circumstances.

But an inability to communicate risk is not the only outcome of systemic failures in science reporting. There are other, less visible but insidious effects. To return to the PVS example for a moment, in 2006 a Belgian researcher named Steven Laureys, along with Adrian Owen and colleagues from the Medical Research Council Cognition and Brain Sciences Unit in Cambridge, UK, described the case of a woman who had been diagnosed with this condition. Brain scans revealed that when she was asked to imagine playing tennis, the pattern of activation in her brain was similar to that in a healthy person performing the same task. The real message from the findings was that the woman had probably been wrongly diagnosed, but when the paper appeared in *Science*, Laureys was bombarded with calls from journalists, asking questions such as whether this meant that some unconscious patients were capable of playing tennis. When scientists see their findings hyped up, given an unjustifiably sinister slant or simply misrepresented, they become defensive and may be less likely to speak to the science journalists who later come seeking information. The danger, then, is that comment and anecdote will rush in to fill the factual vacuum.

Another outcome of these systemic failures is that the relative importance of stories gets skewed. Science correspondents and editors working for major news outlets sift through piles of press releases and other potential source material each day, using their specialist knowledge to weed out and prioritize the important developments. They may write three or four stories a day, of which – depending on what else is happening in the world – only one makes it into the paper. The top down selection process means that this is not necessarily the one the science journalist considers the most significant. It may be a smaller story that makes better copy and in which, perhaps, the science and/or maths is less complicated. The result is that viewing the coverage retrospectively is a bit like walking into a hall of mirrors. One notices surprising gaps in the record, while other stories loom up far too large. During the 1990s, global warming was probably a victim of the vanishing news effect. More recently, Goldacre argues that not enough attention has been paid to changes in legislation governing intellectual property rights, or to the open-source software movement. On the other hand, I can't help wondering if the reported rise of creationism might not have been a self-fulfilling prophecy created by overly generous media coverage – another example of the power of the press, similar to the way in which media reports can drive the conversion of acute mass hysteria into chronic episodes.

4 Lessons to be learned

So what can we do to improve things? Clearly, editors need to make sure that science stories are handled by science journalists, and that they listen to those specialists when it comes to selecting and prioritizing stories, to ensure balanced coverage. This balance will never be perfect, because science competes for space with politics, sport and show business, and because in the end readers decide what they want to read, but there is room for improvement. First and foremost, science journalists should present the evidence. When there is a genuine divide among scientists as to how that evidence should be interpreted, as in the case of CFS, they must give both sides of the argument. When the evidence speaks for itself, in the case of MMR, they must convey the facts as plainly as possible and avoid relying too heavily on comment. Governments can help by offering advice to media organizations about how, for example, to communicate risk – and some of them already do so. In the United Kingdom, the Science Media Centre at the Royal Institution has set a good example by providing in-depth information to the press on science-related issues.

Scientists need to keep the channels of communication with the media as open as possible, and not to retreat into their ivory towers at the first idiotic question – because the idiotic question may be the one a reader would ask, if he had the opportunity. Such a tolerant attitude will pay off in the long run, because getting a message right in the first place is easier than correcting it later. There may be constraints on their openness – journal embargoes and patent applications, to name but two – but again, even within these constraints there is room for improvement. Scientists also need to present their research in a way that non-scientists can make sense of it – for example, by giving risk comparisons that people can understand by reference to their daily lives. The section of the RS guidelines for scientists and health professionals may prove instructive here,[2] and scientists might also benefit from media training (already offered in some research organizations), so that they don't find themselves out of their depth the first time their research makes a splash.

There are bad journalists, just as there are bad scientists, but we shouldn't lose sight of the fact that what the best examples of each profession have in common far outweighs their differences. Throughout history, scientists have risked their lives for their ideas, just as journalists have risked theirs to protect access to information and freedom of speech. Science worthy of the name is of human interest; it cannot be divorced from life. One researcher from a highly respected scientific institution recently complained to me that after the appearance of his paper in a major journal, his lab was effectively paralyzed for a week because key workers were kept busy fielding press calls. Most serious research institutions have the resources that make it possible to prepare for a situation like that. Even with a few days' notice of publication, you can brief a press officer and organize a press conference. Once you have put such practical measures in place, sit back and enjoy yourself. Isn't it gratifying, after all, to have the world's press vying for your attention? Doesn't it say you have done something important? The Apollo 11 mission was scientifically significant, but that's not why we remember it. We remember Neil Armstrong placing his left foot on the moon and saying, "That's one small step for man, one giant leap for mankind."

Harry W. McConnell, M.D., FRCP(C)
Director, Institute for Sustainable Health, Education and Development
Professor of Neurology and Psychiatry
Griffith University School of Medicine, Queensland, Australia

Prof. Harry McConnell is a Neuropsychiatrist specializing in disability with over 20 years experience in both the clinical and public health aspects of health and disability. He has published five textbooks and worked as a Clinical Editor at BMJ Clinical Evidence. He also has extensive experience as an Editor and Reviewer on many medical journals and a keen interest in Open Access and innovative use of IT to make scientific publishing more available in developing countries. Prof. McConnell has worked extensively with the WHO, World Bank, and other international agencies on the implementation of eHealth programmes in developing countries. He trained in the United States, Canada, New Zealand, and the United Kingdom. Prof. McConnell also has a keen interest in evidence based policies for disability services and in health and disability in developing countries. He is a Consultant Psychiatrist and Professor of Neuropsychiatry at Griffith University School of Medicine.

Ashley Pardy, M.A.
Griffith University School of Medicine
Gold Coast Campus, Queensland, Australia

Ashley Pardy is the project manager of the International Health and Disability Network (IHDN), an NGO dedicated to improving health and disability services in developing countries. Now a full time Ph.D. student at Griffith University in Australia, Ashley started her University education at Queens University in Canada, where she completed her B.A. degree and then continued on to Australia to do her Masters degree in International Relations. Prior to becoming the project manager of the IHDN, Ashley taught English in South Korea and she worked as a volunteer worker throughout countries in Asia, Africa, and South America.

CHAPTER

14

Future Trends in Medical Research Publishing

Harry W. McConnell and Ashley Pardy

There seems to be no study too fragmented, no hypothesis too trivial, no literature too biased or too egotistical, no design too warped, no methodology too bungled, no presentation too inaccurate, too obscure and too contradictory, no analysis too self serving, no argument too circular, no conclusions too trifling or too unjustified, and no grammar and syntax too offensive for a paper to end up in print

Drummond Rennie
Deputy Editor, Journal of the American Medical Association

I can predict anything but the future

Yogi Berra

Medical Publishing is on the verge of dramatic changes. Here we will look at why it needs to change, how it will change, and how research can benefit from the rapidly occurring technological and scientific advances affecting scholarly publishing. Publishing has had a profound impact on research for so long now that medical and scientific research has become almost synonymous with the published peer reviewed research article. We seem to have forgotten that the journal article is not the research itself but merely one means of communicating it. By all accounts, a journal article is an artifact of the – to date – chosen media for making people aware of the results. Research has been confined in this manner for many years, and even to an extent defined by its medium of dissemination, the medical journal.

1 Why does medical publishing need to change?

The process of publication has traditionally been the all important final step in the completion of any respectable research project. Academics tend to put great thought and effort into the consideration of what journal article to submit to as there are many important factors to consider in deciding how ones research will be perceived by the public. Academics need to gauge the impact factor of the journal as well as the journals status, the perceived likelihood of publication, and the speed in which it will be made available to the public. Often times the choice of journal submission becomes just as important as the original research, because many people do evaluate the quality and significance of the research based on where and how it is presented. When a researcher publishes his or her work in a journal with a very high-impact factor, the professional ramifications can be extremely beneficial to the individual, often leading to grant funding opportunities and promotions within a University. These opportunities tend arise due to the transference of perception whereby the prestige associated with a journal is transferred to the research published within it. Simply put, when research is published in an important journal then it too is seen as important. Publication in a low-impact factor non-peer reviewed journal, on the other hand, does not have the ability to generate the professional opportunities that the high-impact factor journal does. Universities and grant donors tend to reward and fund research initiatives that are seen as important because they lend credibility to their institutions and in turn strengthen their own reputations and hence funding perpetuated. This quest for academic prestige has led to the creation of a virtual journal hierarchy with the power to dictate how the quality of an individual's scholarly achievement is seen, by association of their own reputations. This complex ranking system surrounding journal classification is dangerous for two main reasons. Firstly, the high cost associated with publishing research in "core journals" prevents many authors with cutting edge research from availing their important information to the entire audience of scientific and medical professionals. The overwhelming problem is that the sheer quantity of journals today has made it difficult for a person with limited time to gauge the value of an article prior to reading it. In the words of Dr. Joel Dimsdale "How can one possibly keep up to date with so many journals... how do we gauge the intrinsic value of an article before we invest out time in it?"[1] The danger is precisely this, the intrinsic value of a journal and by extension the research within it, is often measured prior to one even subscribing or reading the contents. This ranking mechanism has left a wide array of valuable information isolated from the scientific community and without impact on the field in which they are published. Disturbingly, "more than 50 percent of published articles are not cited at all within 5 years after their publication."[2] Secondly, the journal ranking system has created a situation that has placed librarians and the academic community at the mercy of unscrupulous publishers who can increase the price of their core journals tenfold if they so desire. This situation has been dubbed the "serial pricing crisis" by Solomon who reports that the dramatic rise in the cost of serial publications over the last 30 or 40 years has indirectly caused the commercialization of scholarly publishing.[1] This price increase causes the isolation of research even further. Thus the journal hierarchy coupled with serial pricing crisis has made the decision on where to publish absolutely essential for an academic motivated either by prestige, or more appropriately, the desire to make a scholarly impact.

Once arriving at the journal the research then undergoes the lengthy processes of editorial reviewing, peer reviewing, copy editing, and statistical reviewing. After which time the paper can then go back and forth for revision several times between the author, peer reviewers, and editors. There is only one chance to get the paper "right" and hence suitable for paper publication. This process can take 6 to 12 months before the results of the research finally reach paper publication. The end result is meant to be a tribute to the research for which everyone involved has worked so hard over the years to produce. It is also meant to incur clear benefits for the researcher, examples being University promotion and international prestige. Publication in a high-impact journal has been considered by many of the most important, or even the only way to advance one's academic career. Certainly after such great consideration on the part of the authors, editors and peer reviewers and with many months of peer review and editing, the final result of this research product must be the best possible representation of the research and at least that particular research question addressed by the project must surely represent "Truth." Or does it?

Richard Smith, former editor of *The British Medical Journal*, points out in the conclusion of his recent book *The Trouble with Medical Journals*, the dramatic need for urgent reform in medical journals:

"Medical journals have many problems and need reform. They are over influenced by the pharmaceutical industry, too fond of the mass media, and yet neglectful of patients. The research they contain is hard to interpret and prone to bias, and peer review, the process at the heart of journals and all of science is deeply flawed. It's increasingly apparent that many of the studies journals contain are fraudulent and yet the scientific community has not responded adequately to the problem of fraud. Editors themselves also misbehave. The authors of the studies in journals have often had little to do with the work they are reporting and many have conflicts of interest they have not declared. And the whole business of medical journals is corrupt, because owners are making money from restricting access to important research, most of it funded by public money. All this matters to everybody because journals have a strong influence on their healthcare and lives."[3]

The core of medical publishing is peer review, dubbed the ultimate mechanism for quality control. Peer review has become to be seen as synonymous with quality. To be peer reviewed means that ones research has passed the strictest tests of ones peers and that it has been judged to meet critical quality standards. For many years now peer review has been thought of as the "gold standard" in assessing research. Not only does its reputation preclude that research submitted passes through an objective and reliable process set up by international standards, but that the data within the research has also been analyzed according to appropriate statistical methods for publishing medical research. But is it? Peer review is now being critically looked at and processes for quality control in medical publishing are being assessed objectively.

Only in recent years has the process of peer review been itself subjected to reviewing by peers. The *International Congress of Peer Review and Biomedical Publication* is a conglomerate of academic scholars who have met every 4 years since 1986 to discuss important peer review and publication issues such as open access, publication bias, and quality of journal articles. This year's Congress consisted of 470 participants from 38 different countries around the world.[4] The meeting has allowed for a scientific forum where quality control issues of medical publishing and the problems with peer review can be aired.

However, long before the establishment of this regular forum the problems with peer review were apparent to all and, in a sense, can be classified as the worst kept secret of medical research. Indeed the problems were grossly apparent to everyone involved in the process but these issues were rarely exposed in public forums. Jefferson *et al.* has performed a systematic review of the peer review process, concluding that there is insufficient evidence to give a reliable form of quality control in publishing.[5] The problems with peer review are twofold: reviewer misconduct and review bias.

Reviewer misconduct does occur and can be difficult to detect. Peer reviewers are chosen for their prominence in the field related to the specific research being published. As a result, peer reviewers are often colleagues of, and frequently competitors of, the authors of a given paper. However, the conflicts of interest and personal relationships of peer reviewers are not published as are those of the authors. Indeed peer review is in most journals a closed process so that not even the names of the peer reviewers are known to the authors or to the readers. Reviewers may for example, choose to steal ideas from their competitors, or even the actual publications. They may also recommend that the articles of their competitors not be published, or delay their publication buy taking a long time to review the article. They may even take credit for their competitor's ideas and discoveries, with complete anonymity. Richard Smith gives some fascinating personal anecdotes of conflict of interest, plagiarism and other forms of such misconduct by peer reviewers in his years with *The British Medical Journal*.[3]

Bias amongst peer reviewers is more common than actual misconduct and takes many forms. *Gender bias* is one well-known and well-documented form of bias that has for many years seen it more difficult for women to be published then men. *The bias against negative studies* that is against studies which show that a given treatment of intervention has no effect is also well documented. This particular form of bias has the tremendous effect of influencing the whole of the medical literature as it is a commonly repeated bias through many journals, to the point that many authors will not even bother to submit their research unless it shows a positive effect. Another common form of bias in the peer review process is called the *Matthew Effect*, analogous to the Hawthorne Effect in education. The Matthew Effect suggests that individuals with prominent names or coming from prestigious institutions are more likely to be published than others.[6] Peters and Ceci took 12 studies from prestigious institutions which have already been published in psychology journals and resubmitted them with minor changes to the titles abstracts and introductions to the same journals. They changed the author's names and the institutions to names of fictitious institutions without the same prestige of the originals. Only three of the journals realized that they had already published the paper and eight of the remaining nine were rejected because of poor quality.[7] These results show clear evidence against authors from less prestigious institutions and more for name and reputation. This form of bias was so well accepted that, in my own formal instruction in the assessment of research articles in postgraduate training, it was taught as the first item on the agenda in assessing the quality of any given published research paper. *Bias against dissent for mainstream theories* is argued as inherent in the process of peer review by many sociologists. The small number of high-impact journals and their associated reviewers offers partial control over high profile "discoveries" and puts these decisions in the hands of a relatively elite part of the establishment.

Some famous peer review failures that have made the lay press include the Jaques Beneviste and Jan Hendril Schon Affairs in *Nature* and *Science*, where the research was published after being peer reviewed but the results could not subsequently be replicated by others. Similarly, was the recent publications in *Science* by Woo-Suk Hwang, where the results of his research were shown to be fraudulent. The Sokal Affair on the other hand, showed a different sort of flaw in the process. A purposeful hoax by physicist Alan Sokal on the journal *Social Text*, submitted a pseudoscientific paper as if real, which was published and thus sparked a widespread debate on scholarly publication ethics.

To understand why we need to change the process of medical research publication, we need to understand first why we publish research at all. The first editor of *The Lancet*, stated in 1823 that the purpose of the journal in its first issue was to "put an end to the mystery and concealment in the world of medicine." This altruistic early view from a medical publisher differs greatly from the view put forward by Richard Smith on why publishers publish: "To Balkanize Research to make it difficult to do systematic reviews essential in modern research," citing the approximately $US 12 billion profit of

Reed-Elsevier over the past 5 years as evidence.[3] But then why do researchers publish? Harnad surveyed researchers and found approximately 90 percent published to communicate results to peers and 80 percent to advance their careers, with personal prestige and gaining funding also highly cited.[3] Personal financial reward was cited as a reason in only about 10 percent of those surveyed.[3] The motivation of the content producers is thus vastly different than the commercial publishers who capitalize on the market. The market for clinicians wanting to keep up, to answer specific questions on specific patients, or to gain continuing professional development credits, is ever increasing despite the fact that the number of different media they now have to choose from is increasing exponentially. With all the choices becoming available to the reader and to the researcher to submit, commercial publishers need to rethink their business plans to include Wakely's original motivation in establishing the Lancet which was to take into account the needs of researchers and changing scene of media, and offer something that can be seen as real value to the reader.

2 What are the major trends affecting medical publishing now?

There is no question that technology is out pacing innovation in medical publishing, and any commercial publisher who wants to stay in the game needs to be aware of what the major technological trends are. A few current technological influences on medical publishing are given below.

2.1 Open access

- The Budapest Open Access Initiative was started in December of 2001 the Open Society Institute (OSI) held a meeting in Budapest to discuss the international effort to make research in all academic fields freely accessible over the internet.[5] In this meeting researchers and academics from a broad spectrum of disciplines explored the most cost effective and productive means of serving the interests of researchers, institutions, and the public, to make open access both affordable and accessible.[8] The result was a document signed and individuals and institutions from around the world with a commitment to advancing scholarly knowledge. The Budapest Open Access Initiative has been described as "a statement of principle, a statement of strategy, and a statement of commitment."[8]
- Open Source and the GNU license: The General Public License is the most popular example of a copy left license. Under the copy left premise, the GNU is used to grant computer program users the rights to free software definition even when the work has been previously altered.[9]
- The Public Knowledge Project is a Research initiative started by Simon Frasier University and the University of British Colombia which is designed

to improve the scholarly and public quality of academic research through the development of free open source software. The PKP-developed software is used primarily for management, publishing and indexing of journals and video recorded conferences.[10]

2.2 Making archives unique and accessible

The Digital Object Identifier (DOI) has been described as a "permanent identifier given to a document, which is not related to its current location."[11] DOI's are commonly used to give an article a unique numerical identification that can be used by anyone to locate the specific details in a paper. In this manner a DOI functions as a permalink which does not change over time even if the article has been digitally relocated.[11]

2.3 Online authorship collaboration

The Wiki is defined as a "piece of server software that allows users to freely create and edit Web page content using any Web browser. Wiki supports hyperlinks and has a simple text syntax for creating new pages and cross-links between internal pages on the fly."[12] A Wiki is unique because the content and the organization of contributions can be changed or edited by the public. This concept of open editing is a democratic use of the Web that supports content composition by non-technical users.[12]

2.4 Push technologies

Real Simple Syndication (RSS) is a group of Web-feed formats which is used to publish frequently updated content such as news articles and blogs entries.[13] An RSS document contains either a summary of content from an associated Web site or the full text article. RSS "makes it possible for people to keep up with their favorite Web sites in an automated manner that's easier than checking them manually."[13]

2.5 Web bibliometrics

Represent new measures of the impact of a scientific publication in the ways of measuring impact (e.g., citations are no longer relevant in the digital age). The impact of a publication will be measured by a whole new set of measures which take into account "hits", downloads, citations, inclusion in blogs, contributions to paradigm shifts in ideas, and measures of collaboration, influence and input based on "fuzzy logic" and technological measurements relating to the use of study or thesis.

2.6 Social networking

Web 2.0 refers to a second generation of Web-based communities and hosting services such as Wikis and Web blogs which are designed to facilitate sharing and collaboration between users.[14] Web 2.0 alludes to an improvement of the World Wide Web with technological additions that allow for cites such as ebay and gmail that provide an improvement over read only sites.[14]

2.7 New paradigms for copyright shifts

The Copy Left Movement includes the practice of using copyright laws to remove restrictions on distributing copied and modified versions of a work for others to access.[15] Copy left is essentially the opposite of copyright. It is a form of licensing and is mainly used to modify copyrights for music, documents and computer software. Through a copy left license an author can give the public permission to reproduce, copy, distribute or alter his or her work as they see fit.

At the same time, there are some critical medical and scientific trends affecting medical publishing including Evidence-Based Medicine, new methods of peer review, for example Open Peer Review and Commentary, Naboj Dynamical Peer Review, Preprints, Patient empowerment and seeing patients as partners in health care.

Recognition of these trends by a handful has shown a few leaders in the online publishing game as shown below.

Wikipedia

Wikipedia has been rated one of the top 10 most visited Web sites in the world.[16] The reason for its popularity lies in the projects open source software platform which is custom designed to provide the user with access to a large database of encyclopedic knowledge free of charge. The unique feature of Wikipedia is that anyone can edit or contribute to an article using a wiki markup language. This ability to edit information has made Wikipedia the largest collaborative multilingual encyclopedia in the world. Currently Wikipedia is featured in 253 language editions and consists of millions of articles from a vast variety of disciplines.[16]

Faculty of 1 000

The Faculty of 1000 is an online research tool that highlights the most recent and significant research available to the scientific community. The Faculty of 1000 is divided into two sister sites: *Faculty of 1000 Biology* and *Faculty of 1000 Medicine*. Each of these sites essentially acts as a filter whereby over 2 300 of the top scientists in biology and medicine give a recommendation of what they deem to be the most interesting and important papers in a given field of research.[17] This methodology has been highly successful because it organizes and evaluates the large mass of literature available to the scientific community. One of the unique features of the Faculty of 1000 Web service is their Hidden Jewels page. This particular feature identifies important articles that might have been overlooked because they were not published in the top 10 most popular scientific journals.[17]

Philica

Philica is an innovative online academic journal that posts publications from any discipline. The revolutionary aspect is that all academic peer reviews are published alongside the actual article, making the peer review process both unique and transparent. The premise underlying this idea is that peer reviews are often very insightful and can be interesting and useful in their own right to the reader.[18] By making these normally hidden articles part of the public domain, Philica lends a new dimension to the peer review process.

BioMed Central

BioMed Central has emerged as an innovative solution to the problem of high cost academic publications which have been severely limiting the circulation of important scientific research. The independent publishing house offers open access publishing which makes major articles available for the public to access free of charge over the internet. With BioMed Central the high cost of publication traditionally born by the reader is absorbed by the author who is required to pay an arrangement fee of approximately US $500.[19] Currently BioMed Central features about 180 open access journals and offers a rapid peer review and citation tracking services.[19]

PubMed Central

Similar to BioMed Central, PubMed Central is oriented toward providing the public with unrestricted access to scientific journals. PubMed Central however is more of a freely accessible archive of biomedical and life science articles derived primarily from research sponsored by National Institute of Health (NIH). PubMed Central is an innovative method of archiving literature whereby the traditional concept of a library is replaced by a digital repository capable of staying current and reaching vast amounts of people irrespective of geography.[20] With PubMed Central one can quickly search a complete body of full text articles and locate relevant information regardless of where it originated.

Public Library of Science

The Public Library of Science (PLoS) is an open access publishing library that offers free access and redistribution rights to complete journal and scientific articles. PLoS is a pioneer in the field of biomedical publishing because it offers a fast and cost effective way of getting articles into circulation. As with BioMed Central, PLoS charges authors an arrangement fee to cover publishing costs rather than charging the public who wants to access the articles.[21] The unique feature of PLoS is that it offers policy, student and learning forums which gives policy makers the chance to debate health care issues, medical students the opportunity to give their opinions on global health issues, and teachers the chance to access a vast array of clinical-based scenarios constructed by PLoS for teaching purposes.[21]

ArXive

ArXive is primarily used as a library for preprints of scientific papers in the fields of mathematics, physics, computer science, and quantitative biology. This preprint archive can be accessed online and holds more than 423 000 eprints of scientific literature.[22] ArXive was one of the instrumental precipitating factors leading to the open access movement, which has seen scientific publishing move toward providing the public with free access to scientific research rather than having to pay exorbant prices to have journal access.[22] The archives success in preserving preprints has lead to the site gaining complete access to virtually all papers published in the fields of mathematics and physics.

Google Scholar

Google Scholar can be characterized as an interactive search engine whereby scholars can access a vast array of literature, peer review papers, books, abstracts and technical reports. The search engine enables people to find scholarly information in open and non-open access journals. Google Scholar is an innovative way of accessing scholarly research as it makes a wide array of knowledge easy to retrieve based on a keyword search that organizes and prioritizes reports based on a guarded relevance algorithm.[8] Relevance ranking in Google Scholar goes beyond the normal criteria and takes into account a variety of factors such as the prestige of the publisher, the author's articles, and frequency in which the papers have been cited.[23]

HINARI

The HINARI research initiative promoted by the World Health Organization offers universities, medical schools, and hospitals in low-income countries both free and low cost access to online medical databases and journals. HINARI is different from most other open access journals as it targeted toward promoting research and bettering health care practices in over 109 developing countries around the world.[24] The goal of HINARI is to end information isolation by providing health care professionals in developing countries with contacts to over 3 750 nursing, medicine, and social science journals.[24] This free digital access has been touted as an inhibitor of the millennium developments goals as it has facilitated the easy access of valuable medical information used to save lives.

World Association of Medical Editors

The World Association of Medical Editors (WAME) is a global organization designed to facilitate cooperation and collaboration between the editors of peer review medical journals. The organization is primarily virtually based however, meetings are held annually whereby a global constituent of medical editors converge to discuss issues ranging from plagiarism to peer review ethics. Currently there are 1 158 biomedical editors representing 721 journals in over 80 countries around the world actively involved with the organization.[25] The innovative ideas behind WAME have lead to the creation of atmosphere in which editors from around the world can interact together as colleagues and can contact each other via a listserv which is accessible online, for information or assistance with any problems they might encounter.[25] Furthermore the association is responsible for appointing an Ethics Committee which is responsible for developing ethical policies pertaining to medical journals and Web resources.

The Cochrane Collaboration

A group of colleagues searching for evidence on every medical topic world-wide, the Cochrane Collaboration puts together the best in systematic reviews in a truly innovative fashion with a unique form of working together that inspires the best people in evidence-based medicine toward the quest for finding and analyzing all the randomized-controlled trials on any topic and making these results available online.

3 So, what does the future hold for medical publishing?

Clearly the current trends all emphasize *Open Access, Open Peer Review Process, Open Archives* plus multiple and interactive means of disseminating medical research. Indeed, if the main reason for a researcher to publish is to communicate to his or her peers and the main measure of that to date is citations, then such open online access is essential to further research. Harnard and Johnson have shown that an online article receives approximately 336 percent more citations compared to off line publications.[26] Also, the future of medical publishing will thus facilitate participation in the funding, design, production and dissemination of research. It will not be a static form of dissemination but rather the start of an ongoing discussion involving clinicians, patients, caregivers, policy makers and all stakeholders. In the future patients will be involved in all aspects of research including prioritizing of the use of research funds and will become partners in research production with academics.

4 Why is this future desirable?

I am reminded of the quote by Rosen who draws the likely conclusion that giving patents access to medical knowledge does not necessarily represent a danger to medical authority, "Do Net-surfing patients stop trusting their doctors? No, but they are less likely to be overawed."[27] The same conclusions can be drawn about medical publishing, just because people will have access to electronic journals does not mean that they will stop reading printed journals, only that they will be less likely to be influenced by reputation and journal prestige. Indeed, in this future where scientific publishing is characterized by open access, readers will no longer be enslaved by high subscription prices and publication hierarchies. As it is, the pay for access model of publishing alienates large sections of the population, specifically those from developing countries, from accessing literature and contributing their inputs to medical discussion. Open access and electronic publishing fulfills this problem by delivering medical information at cost and henceforth more readily accessible in developing countries, and also allows for greater international participation in medical discussions through such technologies as Web 2.0. Overall, in this new future medical knowledge can be transferred quicker and easier from those who have it, to those who need it.

In *The Structure of Scientific Revolutions* published in 1962, Thomas Kuhn put forward the concept that true scientific knowledge does not advance as a linear increase in understanding based on logical models, but rather as a series of revolutions which occur periodically, replacing the old paradigms and resulting in a true paradigm shift. An older scientific order is entirely replaced with a new order, a new way of thinking about a problem. Based on this concept, in order to advance science, publishing must enhance such a process. In other words, publishing will be a means to create new paradigms in scientific knowledge. These paradigm shifts will have the potential to result in much faster and greater advances in medicine than allowed by the current system which seeks to make knowledge available to a select few by subscription or by membership in so called scholarly societies. In light of these advancements, future success will no doubt be measured by Web bibliometrics analyzing one's contribution to a paradigm shift i.e. the true impact and contribution of one's research to making a real impact. These bibliometrics will replace the current system of publish or perish measured by citations in so-called high-impact journals. No longer will researchers be subjected to the commercialization of scholarly publishing which in many cases, dictates the medium of dissemination in which an author relates scientific and medical information to the global community. The new system presents a unique opportunity to enhance scholarly communication and create an environment of unprecedented transparency in research, making fraudulent research virtually impossible because of the degree of communal participation and transparency at every step along the way. We are already able to see the impact that open access has had on delivering quality, peer reviewed, transparent articles. An excellent example is the "boot strapping" process detailed by Larry Sanger, "Wikipedia articles...are ever improving...Eventually, Wikipedia's articles will far exceed the level of quality that would lead a reviewer to accept them. This is possible because the Wikipedia is, again, ever improving and because the barriers to contribution are minimal."[28–29] This improvement of publishing quality will raise the standard of published articles and thus the value of research material. Open access, open archives, open editorial review, open peer review will be just the beginning. The new system will make open access to original data and collaboration in unique ways not yet envisioned the norm in doing research; data mining will become a routine method of research in medicine and patients will act as partners in producing the protocols. *Publish or Perish* will be replaced by *Impact or Implode.*

Nevertheless, it must me acknowledged that all of these positive outcomes predicted for the future do not come without fear and criticism. Some of the most popular fears expressed as we move more from hard copy to electronic publishing is the long-term accessibility of digital archiving. Anna Okerson is one such academic who has argued that it would be an absolute mistake to rush into destroying a public system that has worked reasonably well for hundreds of years. "Schemes to drive publishers to stop publishing will find 50 ways to backfire, ways that we cannot now fully imagine."[1] The risks inherent in destroying hard copy libraries in favor of digital ones need not be solely focused upon as a deterrent to electronic archiving. Again we can refer back to Rosen's analogy, just because we have something new does not mean that we have to abandon the old. While digital archiving seems to be the mainstream of the future, that does not necessitate that hard copy libraries should be destroyed. Why not have both? It would be perilous to deny the benefit of having a tangible hard copy that you can hold in your hands and turn the pages. Indeed, electronic publishing should exist along side printed copies until digital archiving has stood the ultimate test, the test of time. As Dr. Dimsdale points out, "this dual offering via print and electronic publishing calls to mind the flourishing of both movie theaters and videotape recordings. Both businesses appeal to slightly different markets, and the result has been very beneficial for the film industry."[2] Both printing and electronic publishing meet different needs in an increasingly expanding market. Their presence alongside one another should be interpreted as a compliment rather than a competition.

Given these publishing trends, one can hypothesize that in the distant future we will see the current concept of authorship disappear and the production of research and its dissemination become a process – not a single event marked by publication in a paper journal. There will thusly be many contributors to a given paradigm shift, rather than a single author per se. Currently in some universities there is actually a disincentive to collaborate. A single author publication counts much more in assessing a researcher's academic status than from a multi-author collaboration. The logic of such a system defies everything we understand about making great discoveries in science. In the future, current rights of authorship will be overhauled and collaboration will be encouraged, not discouraged. The role of commercial publishers, if any, is yet to be determined. One thing is for certain, they will not survive based on a business model restricting access to paying customers and they will need to clearly add value to the researcher's publications above and beyond traditional peer and editorial review. An open source approach to research dissemination will ensure true advancement in scientific acknowledge through real paradigm shifts and important innovative advances. The obvious conclusion is that one way or another the traditional peer reviewed journal will have to evolve.

CHAPTER

15

Conclusion: The Ethics of Scientific Disclosure

Peter J. Snyder and Linda C. Mayes

All science is a social enterprise and as such, communication is at its core. Such communication involves more than the informal or regularly scheduled meetings and discussions among research teams, collaborators, and colleagues. Scientific communication includes formal, national and international meetings in which research findings and discoveries are reported, peer-reviewed publications that disseminate research progress to a broader scientific audience, preliminary reports of findings in applications for new research funding, and progress reports in applications for continued funding. By and large, each of these examples describes communication among fellow scientists in either the same or complementary disciplines – discourse among experts in related fields speaking largely the same language if not the same dialect. When scientists begin to cross disciplines, communicative challenges arise. How can, for example, highly technical scientific specialties such as molecular biology or neuroimaging convey the core implications of their projects or programs to other disciplines, such as psychiatry or pediatrics, for whom those implications may be very relevant? Annual reviews, book chapters, or books for above-average informed scientific professionals are among the venues of transdisciplinary communications. Similar, even more significant challenges (or opportunities) arise when scientists speak about their work to non-scientists in professions such as law, theology, the humanities, to policy makers, and to lay audiences. The broader and more diverse the network of communication, the more often scientists are working through other "communicators" such as journalists or professional writers and the further removed their work is from the detailed, technical reporting of individual experiments or research programs.

Further, the demands of communicating to broader audiences and to editors and writers underscore another feature of social discourse – story telling.

By this, we do not mean creative, fictional accounts but rather the essence of conversation between two or more people that necessarily requires effort to help one person understand the intentions, meanings, and narratives of another. Social story telling has many purposes that are shared by the social networks among scientists and non-scientists – sustaining relationships, engaging others, seeking understanding and/or approval, starting new relationships, asking for assistance or favors. These largely positive motives are also joined by the other equally human and equally social motivations including competitiveness, seduction for the purpose of influencing, even misleading another, campaigning for personal gain or promotion. For the scientist, these very human incentives in social discourse are in the service of furthering their science and linking to other colleagues, enhancing their laboratory's productivity, their own academic career and reputation, their funding support, and influencing policy makers and foundations. No doubt these motivations influence when and how the scientific story is told, that is, what findings are emphasized, which ones are put to the side for further study, how findings are interpreted, and how far reaching are the posited implications.

When the science has a more direct impact on other individuals, as in many branches of the biomedical sciences, another important goal of scientific communication is to improve the medical care of patients, and those patients become key partners in the network of biomedical scientific communication. How biomedical research is communicated to patients may change the individual patient's health care and dramatically influence the standard of practice in any community. Further, when scientists choose to communicate their findings regarding, for example, new treatments for a serious disease or new preventative interventions may be life-altering for many individuals. Findings released too early or too late – or findings that are inevitably evolving given the dynamic nature of much of biomedical science – stand to impact many patients positively or negatively as well as to influence the direction of work and funding in any given field.

As described in the introduction to these essays, the intention of this volume has been to address the many strands of communication among the biomedical scientific social network and the issues that characterize these often very specialized social networks. We have included representatives from many members of the biomedical scientific social chain – scientists themselves from the academy and from industry, science writers, scientific journal editors, ethicists and theologians, historians viewing the evolution of scientific communication, and lawyers in the thicket of the boundary between new science and policy. We believe that the process of biomedical scientific communication is so complex and so at the heart of good – and bad – science that it deserves scrutiny, discussion, and its own scholarship and have set our essayists the charge of laying out the terrain of the issues that scholarship might cover. In the sections that follow, we reflect upon the issues our essayists have identified and on the implications of these issues for the health and future of biomedical research.

1 Discovery and authority

As biomedical scientists, our behavior and professional conduct is molded by many hard years of education, we complete rigorous training in scientific methodology, and we grow comfortable with exploring the boundaries of current understanding. Unlike many professions that require its practitioners to limit their judgment and actions to what is perceived to be factual information, scientists routinely glide along the razor's edge between factual knowledge and new discovery. As such, the self-perceived ability to remain cautious in interpreting data, to recognize the limits of our own knowledge, and our deep appreciation for the awe-inspiring complexity of our natural world and the phenomena that we study, all become part of our professional identity. Most of us are proud of these finely honed intellectual skills, and our beliefs in our own objectivity rest on our adherence to proper scientific methods and the testing of refutable, falsifiable hypotheses. And yet as many of the authors represented in this volume point out, we are all part-and-parcel of our societies and cultures, and we are prone to the same human moral and ethical failings that are characteristic of our species.

In this respect as we emphasize in the premise of our overview, the conduct of science is largely a social pursuit like any other activity that adults are engaged in for the majority of their lives. As with any other practitioner in any other profession or trade, we want to advance our careers, to be promoted within our institutions or to be attracted by enticing career opportunities elsewhere, to provide needed resources to our students and fellows, to complete important work that will be remembered long after we are gone, to provide financially for our families, and occasionally to enjoy lovely vacations in tropical locations. All of these goals are advanced by disseminating the stories about our work, our experiments, and our discoveries – to our colleagues in the peer-reviewed literature, and/or to the public, our employers, and to funding agencies through all types of media outlets. Our reputations, our authority, and ultimately our career and economic success depends on our ability to tell the stories of our own research accomplishments.

All three central figures described by Profs. Warner, Snyder, and Spencer in their essays – from Louis Pasteur to J.M.R. Delgado, and more recently to Ignacio Madrazo – share an important common belief; that is, these were scientists on the verge of making breakthrough discoveries to care for those afflicted by devastating diseases or conditions. For all three of these important figures, this belief was entirely justified and appropriate. All three were honorable, respected, and talented individuals, and each believed that he was practicing as a scientist with the best interests of his patients, and of the society in which he lived, at heart. For example, in a lecture and published essay to a large audience of public school teachers, in the late 1960s, Delgado closed his speech by stating "I would like to [ask] your cooperation in spreading the message to your students that the mind is no longer considered an obscure metaphysical entity, but a concrete experimental reality with functions which may be evoked, inhibited, recorded and analyzed ... Indeed, knowledge of the human mind may be decisive for our pursuit of happiness and for the very existence of mankind."

This notion that Delgado was engaged as an explorer, an adventurer, in making new discoveries that might be decisive in shaping the future – and even the "very existence" – of mankind, is certainly a grandiose vision of one's own work! Nonetheless, statements such as this one reappear throughout Delgado's written publications in the late 1960s and 1970s, and hence, it is probably the case that he firmly believed this to be an accurate reflection of the importance of his work. Likewise, the history of biomedical science is replete with other figures (including, but hardly limited to Pasteur and Madrazo) who believed, either justly or unjustly, that they were making new discoveries that might alter the course of humankind, or at least to cure major causes of suffering and disease. In all of these cases, as an individual works on the cusp of major discovery, we find repeated examples of how easy it is to cross ethical boundaries in reporting research successes to the public or to one's peers. Indeed, we might argue that it is human nature to become caught up in our own excitement, self-confidence and certainty in our knowledge, even grandiosity, and in so doing, we loose sight of more reasoned, humility tempered judgment. Importantly, in most of these cases it is very difficult to assign blame or guilt for such ethical lapses in judgment, as in most cases the person involved tried to behave on an honorable, ethical manner. It is with the advantage of hindsight that we are able to conduct "historical autopsies" on a case-by-case basis, to learn how we might improve our conduct as scientists and as members of our societies.

2 Emotions and human foibles

As biomedical science has become more complex with advancing technology, our "case examples" has also become more nuanced and multifaceted. They include interactions between science and industry, consumer advocacy groups, ethical review boards, and better informed science writers. Peter Fonagy gives the details of the concerning reports regarding increased suicidality among adolescents treated with certain selective serotonin reuptake inhibitors (SSRIs). This debate came to the forefront in 2005 with reports from the US and UK regulatory agencies, but the association had been suggested as early as 2003. It was an emotionally charged issue because the concerns involved adolescents struggling with depression, a problem especially difficult to treat adequately in this age group and difficult for adults often to acknowledge. The issue of SSRIs for adolescent depression is further complicated by the fact that depression often spontaneously remits in young boys and girls, an observation that raises questions generally about the use of psychotropic drugs for this disorder in this age group. The debate regarding SSRI use in adolescents also touched on the larger concern regarding accurate reporting of the findings in psychopharmacological treatment trials and the suspicion sometimes fueled by media reports of selected reporting by the pharmaceutical companies, especially when the results of the trial were not altogether favorable for the drug in question. For example, in 2004, Eliot Spitzer, the then attorney general of New York, sued GlaxoSmithKline for suppressing data suggesting that Paxil increased suicidal ideation in children. The company settled the case by establishing a restitution fund, paying New York State for the cost of the investigation, and agreeing to make the results of the clinical trial public.

The key observation stressed by Fonagy that sparked the debate was an apparent doubling of the rates of adolescent suicide or suicidal ideation across a range of therapeutic trials. This observation set off mounting concern over months both about selective reporting in randomized trials and about pharmaceutical companies' active suppression of data regarding adverse events. The media coverage was so pervasive and the public response so intense as to stimulate fictional spin-offs in popular television dramas. For example, in one episode of *Law and Order* the suicides of two college students on antidepressants led to the prosecution of the CEO of the pharmaceutical company for murder, because he knowingly continued to pursue development of a drug that clinical studies had demonstrated created a high risk of suicide in its users[1] (an episode probably following on the GlaxoSmithKline case cited above). As attention in the media intensified, the biomedical community responded dramatically. The Food and Drug Administration (FDA) placed a black box warning on these drugs regarding their use with depressed adolescents. Within a very short time, prescribing patterns by child psychiatrists across the country changed, and parents often demanded that their child be withdrawn from all psychotropic medications.

However, as Fonagy points out, with careful examination, the story was and is more complex. While it did seem true that there was an increase in suicidal adverse events in the first weeks after initiating SSRI treatment, subsequent meta-analyses continued to show an odds ratio for suicidal adverse events not as high as originally reported though surely elevated. Further reviews also showed that the majority of SSRIs except fluoxetine were not effective generally for the treatment of adolescent depression. And subsequent analyses showed that medication benefits were most apparent when used in concert with psychosocial treatments. Thus, under calmer scrutiny, there were far more reassuring data regarding the risk–benefit ratio of fluoxetine specifically for adolescent depression. In Fonagy's appraisal, this is one example in which more thorough science was effective in reshaping a debate. The initial storm led to heightened awareness which, in turn, stimulated further investigation showing that the problem was real and worse with some SSRIs and not others, and that accompanying psychosocial treatments reduced the suicidality associated with fluoxetine. Randomized trials, careful meta-analyses, more sophisticated statistical models came together to provide a more precise, practice-oriented response to an initially very concerning report. Fonagy suggests that the initial controversy that could have led to a sustained backlash against psychopharmacological treatments with adolescents had three sources: (1) the close association between industry and science leading to perceptions of undue influence, (2) the strong but often misplaced faith in evidence-based practice based solely on data versus evidence-based practice that integrates accumulated clinical experience, and (3) the failure to take into account epidemiological data regarding the generally positive effect of SSRIs in reducing overall rates of suicide. The third point is an especially complicated one for presentations in the media and for policy makers for it requires balancing questions of risk for an individual versus benefit both for that individual and for society.

The story of SSRIs for adolescent depression is also an example in which the reporting in the media, while sometimes exaggerated, did alert the scientific community to a potential problem. However, the initial press releases and media reports were surely dramatic. How such drama may impact the general public and be an expression of the motivation of both scientist and reporter alike is the subject of the essay by Smith, Klink, and Landwirth (Chapter 6). These authors provide the first exposition in this volume of the competing forces that complicate both the accomplishment of biomedical research and the reporting of that research. For scientists, issues such as the need for funding may go hand in hand with a desire for notoriety (which may impact their success at obtaining funding). These complementary issues may also impact how scientists describe their research to various media sources. Are they more cautious or more expansive? Scientists may be tempted to exaggerate their claims in a news release in the spirit of making a story more accessible or more interesting. For example, a story may be slanted toward the direst complications of a given medication such as the concern regarding increased suicidality among depressed adolescents on certain antidepressants and then on how such data were not made readily disseminated by industry sponsors of the drug-field trials. While in one part a statement of fact, as Fonagy also illustrates, and as was also stressed as well by Spinney in Chapter 13, what is left out of the story is the relative risk for the serious or adverse event compared to the risk of not receiving treatment.

Taking as their "text" two reports on human stem cell research, Smith, Klink, and Landwirth point out that once an issue is framed by the media (based often on initial conversations with scientists), it can be very difficult to reframe the debate for the public, policy makers, or interest groups. It is not that framing toward direr or more engaging aspects of a story does not sometimes serve important purposes (as was true in the example of SSRIs for adolescent depression), but the debate may be inaccurately distorted and very hard to shift once started. The question of stem cell research is a case in point. As soon as the debate was framed as findings that did not "harm" a fetus, the potential scientific value of a method for harvesting blastomeres was caught up in the religious and cultural storm surrounding use of fetal tissue. Similarly, the use of stem cells as one method for spinal motor neuron regeneration was captured in the press releases as offering clear hope for paraplegics with little attention to the absence of human data and also to the need for replication in a series of preclinical experiments.

In both these instances, scientists and reporters alike were caught up in either imaginative or distorted claims regarding the relevance of their work. Each was working on an engaging story that increased the visibility of the scientist and the standing of the writer. Each served the other's needs. Smith and colleagues ask what's wrong with hype? Do exaggerated or distorted claims regarding the implications of scientific findings pose ethical quandaries? Who bears responsibility for the distortions? Distorted reporting in press releases or news stories run the risk of discrediting both the scientist and the reporter and hence may seriously undermine the public's trust in the authority and accountability of biomedical science. Public wariness regarding the accountability and veracity of reports of scientific advances in turn negatively impacts public willingness to support biomedical research. Further, it is not just public mistrust engendered by unnecessary scientific exaggeration that is regrettable, but also overstated claims may mislead physicians to begin or end treatments for their patients on the basis of incomplete information. Responsible care – and as Smith and colleagues add, responsible public policy – depends on well-informed clinicians, citizens, and public officials.

Importantly, journalists choose their stories often from press releases that scientists may have a hand in developing with their public relations offices or by the scientific journal itself. Indeed, Smith and colleagues cite an amazing 84 percent of reports on journal articles referred to as *in press* releases were initiated by the editors of the scientific journal. Clearly, the first nidus of accountability is with the relationship between the scientist and the publishing journal. Both the editor and the scientist need to agree on the boundaries of what can and cannot be claimed in a press release. The reporter following up on the press release also has a responsibility to dig deeper, to be more informed, and to check his or her sources. A breakthrough that seems too good to be true may indeed be so and reporters need also to seek the opinions and comments of other scientists in the field. As will be a theme in other chapters in this volume, accountability and responsibility depends on scientists understanding more about the journalists' craft and vice versa, science writers becoming better informed about the science they are trying to elaborate on for a more general audience.

But Smith and colleagues also bring to the forefront a more complex debate between the value of free press and free debate and a more regulated oversight of scientific reporting. Can free debate and discourse really be counted upon to deflate initial hype and ultimately gravitate toward a balanced, veridical, and complex accounting of the scientific issue at hand? Smith and colleagues argue that human nature being what it is, we cannot reasonably depend on either journalists or scientists not to succumb to the benefits of hype for either of their professions. But at the same time, they argue that both have a professional moral or ethical responsibility to society to report the truth and that this is tacit agreement that both need to make in their efforts to disseminate science. Each needs to agree on standards of self-regulation based on an understanding of moral responsibility to convey science accurately no matter how complex or how incremental. It is a daunting task far easier to say than to do, and this was the theme that others in this volume, notably the Very Reverend Wesley Carr (Chapter 9), revisited several many times.

3 Competing interests and conflicts in scientific communication

As several of the chapters in this volume suggest, there are a wide variety of potential sources of secondary gain that might affect whether and when we either consciously or unconsciously stray from known scientific fact, simplify a story to make it more interesting and compelling, withhold disclosure of all available and relevant study results, or merely embellish a point that we believe with "near certainty" to be valid. Any of the potential sources of secondary gain, ranging from positioning oneself for promotion and tenure, to gaining added respect amongst colleagues, or in response to direct or indirect economic incentives, lead to core conflicts of interest. With respect to financial incentives, as an example, the broad question raised by Prof. Kassirer (Chapter 7) is whether scientists can be objective if they stand to gain financially from the disclosure of study results or a medical opinion, and whether full disclosure of financial ties in any way mitigates this problem. Kassirer's firm answer to this question is absolutely not, and he echo's the voices of other contemporary academics who have studied this issue, such as his colleague at Tufts University, Prof. Sheldon Krimsky.[2] Krimsky has likewise noted that as "commerce and academia become entwined, 'disinterestedness' is on the wane throughout science."[3] In general, however, scientists tend to dismiss the idea that their behavior might be either subtly or obviously altered by the presence of related financial or funding ties. In fact, it has even been argued that "disinterestedness is no longer viable or necessary to protect scientific objectivity."[3] Yet, as Kassirer clearly shows in his essay, well-controlled quantitative studies confirm the presence of a "funding effect." This has been shown in a variety of way, including a 2003 article in the *Journal of the American Medical Association* showing that published randomized clinical trials were more likely to favor an intervention if the study was funded by for-profit corporations (i.e., the manufacturers of the therapeutics being studied).[3,4]

It was on the basis of such positive clinical trials, published when the new class of SSRIs was introduced as antidepressant medications in the 1980s, that they earned the public's trust as being safe, effective, and non-addictive. As Prof. Fonagy also suggests in his essay (Chapter 5), this reputation was and is fiercely disputed, with the possibility that safety risks to children were either not disclosed promptly by one or more drug manufacturers, or that the clinical trials data required to assess such risks were not forthcoming. Indeed, access to clinical trials data, from the pharmaceutical companies who conduct most all of the research prior to a new drug approval, has long been difficult – if not impossible – to obtain. The problem is a serious one for two compelling reasons. First, the "under-reporting of clinical research is biased and can be lethal."[5] That is, under-reporting of clinical trials results, by definition, is selective reporting – and there is ample evidence to show that this almost always leads to an under-reporting of potential adverse events or complications with therapy. No interested party would intentionally under-report good news about a marketed drug!

Secondly, the under-reporting of negative or contradictory studies may actually impede the progress of biomedical advancement, by making it impossible to compare-and-contrast across positive and negative studies in order to learn how to improve on the methodology of the clinical studies we design.[6,7] Biased under-reporting of clinical trials has been recognized as a serious problem for at least the past three decades. For example, in 1980 a Finnish scientist, Dr. Elina Hemminki, published a paper showing that among recently approved psychotropic drugs, those with more side effects were the topic of fewer published peer-reviewed medical articles; whereas those drugs with few side effects had more clinical trials data published for public access.[5,8]

As a response to this "funding effect" noted by Kassirer, Fonagy, and others noted above, many journals, scientific societies, and universities are espousing the effectiveness of full disclosure, or transparency, as a cure for this serious problem. As Dr. Kassirer explains in his essay, although the practice of full disclosure is seen by many as the cure that we need, it provides little more than palliative treatment for a core problem that cuts to the heart of public trust in biomedical science. As Krimsky explains, "disclosure merely legitimatizes the practice of mixing commerce with science and implicitly accepts the decline of disinterestedness."[3] Given the social and economic structures of medicine and research today, what can we do to really confront this problem? Krimsky would argue that the roles of "those who produce scientific knowledge should be kept separate from those who stand to financially benefit from it."[3] Although Kassirer seems to agree with this as a goal and as an ultimate solution, he clearly sees the serious political, social, legal, and funding complexities that make this goal an exceptionally difficult one to attain in an absolute sense.

Many of the concerns raised by Kassirer are not refuted by Dr. Declan Doogan in his subsequent essay (Chapter 8). Doogan, a very well respected senior clinical leader in the pharmaceutical industry, who notably led the clinical development of an SSRI (sertraline) for Pfizer Inc., acknowledges that many of the concerns reviewed above are legitimate. However, Doogan reminds us that, for better or worse, as a result of changes in Federal laws and regulations in the early 1980s, industry funding of research has become the lifeblood of academic medical institutions. In fact, in an era of declining Federal (National Institutes of Health, NIH) support for biomedical research, partly as a result of our government's funding a protracted and tremendously expensive overseas war effort, academic medical centers and universities have become frankly dependent on corporate sponsorship of research. Moreover, with declines in reimbursement rates for inpatient expenses, and increases in the costs of keeping a hospital staffed, equipped, and running, the medical centers increasingly look to overhead charges from research grants and corporate sponsorships, just to balance annual hospital budgets.

There is no doubt that commercial funding of research acts to shape the very questions that researchers choose to ask in their research, the public advice that many scientists offer, and ultimately many of the medical decisions that they or their colleagues make in caring for patients.[9] The massive infusion of private funding for research and development has changed the way many – if not most – universities conduct their business, with many institutions and their faculty enjoying equity partnerships in biomedical and technology companies whose research they are charged with monitoring for ethical compliance. Still, Doogan warns us not to forget that this imperfect system has, over the past several decades, nonetheless led to some of the most dramatic advancements in medical discovery and care that humankind has ever witnessed. Doogan's essay provides a balance to the legitimate and worrisome issues raised by Kassirer, and both together remind us that we are facing a massively complex societal problem for which there are no easy answers.

Virtually all modern university medical centers are dependent on annual research funding in order to support, at least in part, their clinical missions and services that are exceedingly expensive to maintain (e.g., hospital emergency departments). Just as an example, the Yale University School of Medicine receives 65 percent of the total amount of any grant awarded by the NIH to its faculty, over-and-above the actual grant amount, to support indirect costs. The Haskins Laboratories, which is a quasi-independent research center that is jointly administered by Yale and the University of Connecticut, currently charges 75 percent of a grant award, on top of the actual award itself, to the NIH in order to cover operational expenses. These indirect costs cover all institutional expenses ranging from keeping the lights on in the building to supporting functions of the institution that are important, but not profitable. On average, if an institution brings in $70 million per year in grant awards (for direct research expenses), it also receives about $40 million per year to cover indirect costs. This latter sum represents a critically important source of revenue to medical centers that can barely cover the rapidly rising expenses for operating hospitals and clinics, in face of contracting reimbursements from insurance companies. Dr. Doogan makes the case that all medical centers have become dependent on this revenue stream, and with cyclical and unpredictable funding from the NIH (depending on US budget deficits and tax revenue being diverted to other causes, such as support of an unpopular war rather than on social services and biomedical research), medical centers are increasingly looking to the private sector for steady continuance of research funding.

The reality of this situation cannot be ignored, but neither can the very serious and disturbing trends described by Prof. Kassirer, all of which serve to undermine the very integrity and veracity of contemporary biomedical research. Kassirer offers a clear description of the dangers that we are facing in this regard, but again, there are no solutions that would be simple to implement without concurrent massive changes in laws, regulations, and national priorities that would ripple across our social structure. A small but important initial stop-gap measure is raised by the Very Reverend Wesley Carr, in his essay (Chapter 9), and that is that each of us must be held to be personally responsible for what we say and publish. This seems on the surface to be a simplistic, almost trite, notion, and yet all too often we fail in terms of personal accountability.

4 Scientific responsibility and authority: Who is ultimately accountable?

One of the prevailing metaphors of science is the lone discoverer, working against all odds in the service of new knowledge that radically changes medical treatment, benefits countless children and adults, and ushers in a new and healthier social order. He or she is a hero and singled out for individual recognition. John Warner in Chapter 2 touches on the metaphor of the scientist as intrepid explorer as contrasted with the reality of science as incremental, breakthroughs as rare events, and discovery as a team effort. Contemporary biomedical science offers among the most dramatic examples of "team science" for with technical advances, it is increasingly difficult for any individual investigator to be the master of all that is required to accomplish any biomedical research. Neuroimaging initiatives require physicists, engineers, and mathematicians to join the team of diagnostic radiologists, experimental psychologists, and clinicians. Human genetics involves computational and molecular biologists along with internists or pediatricians. Medication trials for new cancer drugs may require biostatisticians with expertise in designing randomized trials, clinicians caring for patients, clinical chemists assessing biological outcome measures, diagnostic radiologists measuring the change in the tumor, surgeons intervening as needed to remove substantial tumor masses, and so on.

The increasing necessity for team science in biomedical research is balanced against the issues of scientific responsibility and authority. When so many investigators are members of a research team, who is ultimately accountable for the research findings? Who speaks with the authority of both knowledge and personal responsibility? Of course, accountability relates to issues of disclosure as outlined by Kassirer in Chapter 7 and to who bears responsibility for communicating science accurately to the public as outlined by John Warner and Smith, Klink, and Landwirth in Chapters 6. Questions of accountability and responsibility are also relevant to the thorny issues of how institutional promotion committees recognize individual contribution among a large team of authors contributing to a published work. Is an individual listed as second or third author who nonetheless designed the computational algorithms for analyzing the imaging data less authoritative as a contributor compared to the first author whose is the principle investigator on the grant funding the study and who was the primary instigator of the research effort? Is one contribution weighted more or less heavily in the academic promotional process? We shall revisit these questions after first addressing in more detail who among scientific teams rightfully bears authority for how and when biomedical research is communicated.

The Very Reverend Wesley Carr brings us to the theme of authority and accountability from a religious perspective. As Dr. Carr reminds us, *Magna est veritas et praevalebit* or "Truth is mighty and will prevail a bit." The terrain between science and religion becomes less of a contemporary Maginot Line when we bring ourselves to ask the nature of truth and authority as it relates to our abiding faith in objectivity of scientific discovery. Dr. Carr shows us that the abiding concern of both scientists and theologians is the nature of faith and of authority. For both, faith is given voice in the efforts toward discovery (those of science or of the soul) and the power of truth that may emerge from those discoveries, and both science and religion are concerned with who has the authority to speak about such truth. Dr. Carr's informed editing of the Latin verb reminds us that truth is indeed relative, especially in science and that both religion and science struggle with concerns about truth, faith, and authority not just in the abstract but in the here and now, in everyday life. Today's breakthrough is tomorrow's quaint ignorance, and as we have already emphasized, all discovery in biomedical science is incremental and changeable. We only have to look as well at the tremendous individual variability in any patient response to a medication for hypertension, malignancy, heart failure, or depression to realize that when we encourage our patients to trust that the most up to date advances in medical treatment are the best, we accept both the responsibility of our authority and the humbling knowledge that what seems indisputably true now may not be so for our patient or even for all patients after more research is accomplished.

Dr. Carr returns to the theme of scientific story telling – or as he calls it constructing a "fable" as a necessary part of the scientific process. This construction, especially in Dr. Carr's use of the word "fable," inevitably brings up both issues of truth and authority and not just as relates to the scientists but to those who speak on their behalf – the journalists and other media professionals. So strong is the human need for guiding authority that we often delegate our own capacity and responsibility to judge what is true to other sources. Our looking to and for authority is not just reflected in "Scientists say…" or "My doctor told me" but also "I saw it on television (or read it in the newspaper…," "I heard it on the radio…," "I read it on the internet… ." Increasingly, in our technology rich world, we turn to media that is several steps removed from direct social engagement and also often offers a considerably distilled and thus, often distorted, account of scientific work. Dr. Carr also reminds us that when scientists, in an effort to communicate their findings, tell their stories to engage the imagination and faith of others, they are playing with the boundaries of authority. Who really speaks for Delgado's work – the neuroscientist testing his hypothesis with a particular species of bull, or the scientist as an "unarmed matador" risking his life in the ring with the fierce, aggressive "brave" bull? As Dr. Carr shows us, engaging the imagination and faith of our audience in the authority of science exposes the many hidden agendas of the scientific craft – needs for funding, authority granted by notoriety, professional stability – and paradoxically even as the account compels its readers to embrace the promise of the new science, it also undermines the ultimate authority of the scientist and seductively undermines his or her capacity to see the patterns of "truth" amidst the noise of scientific data. Further, the more scientists present themselves or allow themselves to be presented as fearless explorers dissecting questions at the very core of life itself, or as Dr. Carr puts it "playing God," the more they both frighten society and ultimately undermine their believability and thus, their authority.

Just as media incautiously used threatens scientific authority and truth, so does the many membered scientific team also potentially threaten those principles of responsibility and accountability so at the core of intellectual discovery. The more authors on a scientific report, the more diluted the ultimate authority and accountability for the findings in that report. Is it the first author or the first two or three authors who really are accountable? What about in the culture of schools of medicine when the senior or last author is often the principle investigator on the grant or the director of the laboratory? Multi-authored papers are a product of interdisciplinary team science and of a culture in which numbers of publications, at least to some extent, is a metric of both authority and academic success. A 1999 review in the *Lancet* underscoring this trend of multi-authored papers noted how difficult it is to identify the specific contributions of individual authors to the components of research from designing a study to preparing the manuscript.[10] Dr. Carr cautions us to be thoughtful about the implications of such an increasingly common practice and not to accept the all too common explanation "pressure to publish" offered by authors when a publication is called into question. Indeed, Dr. Carr suggests that the very faith in the "objectivity" of science has blinded its practitioners to recognizing the fragility of responsibility and accountability in scientific reporting as currently practiced. He offers practical suggestions regarding limiting the number of authors and having explicit statements about each individual's contribution to that publication. Dr. Carr further suggests that in order not to delegate their authority and their accountability to the media that scientists actively learn how to communicate their findings and their work to a lay audience – that they themselves take the responsibility of creating their "fables," their story to be told. We might also add that there needs to be more debate regarding the all too common practice in biomedical research of including lab directors or department chairpersons as authors even if they have not actively participated in the research or the writing of the manuscript. The case study we cited in our introduction regarding prayer and *in vitro* fertilization offers an illustration of this practice. For example, in a review of three major medical journals with clearly stated policies regarding criteria for authorship, the number of authors per paper was on average was seven and the *Annals of Internal Medicine* had on average 21.5 percent papers with at least one "honorary author," that is, an author who had not made direct contributions to the conduct of the research or to manuscript preparation.[11]

In the concluding section of his essay, Dr. Carr touches on the risk of the politicization of scientific authority especially in debates that are so central to what it means to be human (e.g., stem cells) and to the health of the world we live in and our responsibility to that world (e.g., the climate change debate). Politicization of scientific communication affords another challenge to truth, authority, and accountability taken up by Dr. Ruth Katz in her essay on the "Intersection of Government Science, Politics and Policymaking" (Chapter 11). Dr. Katz reminds us that there are often conflicting authorities in science as the quotation from Dr. Bernadine Healy so starkly reminds us: "Sometimes science needs to take a 'time out' when its goals collide with the moral concerns of the society (page 130)." A "time out" from what – accountability, responsibility, reporting scientific findings regardless of whether or not the findings support the original hypotheses or aims of the investigation? Dr. Katz goes on to meticulously lay out a troubling but all too familiar and well-documented account of how scientific communication can be shaped (or distorted) by funding sources, employers, the political and social milieu, and especially by the dual roles that scientists often find themselves in when they accept jobs in the government – or in industry or university administration. The symbiotic relationship between biomedical science and the NIH among other governmental funding agencies means that scientists are always between the proverbial rock and a hard place, only in many instances these rocks are often ethical ones. In the effort to obtain funding to sustain their laboratories, scientists may shape their work to follow the agendas set forth by the NIH; they may tell their stories in a special or limited way in order to diminish the chances for criticism by NIH supported review committees made up of their peers also in similar situations of needing and seeking government funding. And the research agendas for the NIH are set in part by Congress and by the budget allocations to the individual institutes.

While acknowledging the pressures on scientists from government funding agencies that in part shape science, Dr. Katz underscores the potentially greater and more insidious threat to scientific authority and accountability that comes from the suppressing and misrepresenting of scientific findings in order to better support or not overtly undermine a given administration's broader agendas. Dr. Katz makes use of the case study method and draws her examples from five White House administrations, each with their own particular approach to scientific communication and the interface of science and political agenda. As she illustrates clearly, good science (by whatever criteria) and effective scientific communication is not the sole determinant of administration policy. However, scientists may assert their authority and even redress the threats that authority and accountability outlined by Dr. Carr, the reality is their word is not final when it comes to the interface between government administration and dissemination of work by the nation's scientists. Indeed, many administrations used a range of tactics to deal with science that is not compatible with administration goals or policies – or as Dr. Katz calls it, "uncomfortable science (page 140)."

Her examples include the Carter administration's response to *in vitro* fertilization research that led to a 15-year moratorium on any federally funded work in this area. Another example is the Reagan administration's instructions to its officials not to speak publicly about AIDS and its inadequate support of public health efforts to reduce the spread of HIV. The theme of concern regarding scientific study (or interference depending on the perspective) with conception and fetal health was continued in the first Bush administration in which support for fetal tissue research was effectively cloaked in the emotionally laden debate over a woman's right to an abortion – an especially volatile issue for a major supportive constituency for both Bush administrations. The Clinton administration returned to the HIV–AIDS arena in its refusal to lift the ban on federal funding for needle exchange programs despite compelling scientific evidence of the effectiveness of such programs to reduce the spread of HIV. Unlike the Reagan administration's concern about appearing to be allied with the gay rights constituency, the Clinton administration appeared swayed by concerns that it not appear to be facilitating drug use and addiction. And Katz's fifth case study from the second Bush administration offers numerous examples of active suppression and misrepresentation of research regarding global warming. This administration has carried the politicization of science to more concerning and overtly controlling strategies in its active censoring of information regarding climate change from federal reports and its actively interviewing scientists seeking government jobs for their opinions regarding faith-based initiatives, drug use, needle exchange, stem cell research, and other controversial areas.

Many of Dr. Katz's examples are compellingly drawn from scientists working in government agencies such as the intramural programs in the NIH. But just as compelling are the many instances of active suppression or misrepresentation of findings from scientists working for government-funded defense contractors, as illustrated by the 1986 NASA *Challenger* space shuttle disaster.[12] In this well-documented and tragic case, several engineers employed by the government contractor, Morton-Thiokol, including Mr. Roger Boisjoly (a solid rocket motor seals expert), recommended delaying the launch of the *Challenger* on January 28, 1986, due to data they had indicating a high likelihood of failure of the O-ring seals on the shuttle's solid rocket booster (due to expected alterations in the physical properties of the rubber seals in freezing temperatures). Boisjoly and his colleagues sent 13 charts to NASA, to describe their data and to support their recommendation to delay the flight. However, senior NASA administrators were under direct pressure from the White House to make sure that the launch proceeded, so that President Ronald Reagan would be able to accomplish the public relations *coup* of talking directly to first-time astronaut and school teacher, Ms. Christa McAuliffe, during his annual State of the Union Address to a joint session of Congress, and to the American audience, scheduled to occur 10 hours after lift-off. Under the weight of such political pressure, NASA officials were quick to identify weaknesses or ambiguities in the engineer's charts of data, with one high-level official stating that he was "appalled" by their recommendation – this, despite the fact that this was the only such no-launch recommendation made by Thiokol mission scientists in the history of the shuttle program.[12] In response, the management at Morton-Thiokol pressured its own scientists to back down from their initial recommendations in a gut-wrenching midnight teleconference, and the launch proceeded the next morning with horrific consequences. Roger Boisjoly and his colleagues have now lived with the deep regrets, feelings of guilt, and immense damage to their lives and careers as engineers, for more than 20 years.

One implication of each of these case examples, and especially the ones from the second Bush administration, might be that there is little scientists can do to prevent the misrepresentation of their work or the shaping of necessary funding streams by political agendas. However, Dr. Katz adds to the discussion of accountability and authority represented by the papers in this section by bringing in the dimension of personal accountability and authority. Ultimately scientists have to decide then their own values. They must decide whether or not their own standards of accountability are being sufficiently compromised in their roles either as government employees (or as funding recipients) that they must take a different course of action – speaking out in their behaviors, their writings, or both. They must acknowledge that the myth of "pure science" is just that, a myth that is both naïve and dangerous in that it excuses scientists from becoming more active in defending against inappropriate politicization of research and from struggling with the very real conflicts of interests and moral dilemmas that they are faced with when their work inevitably finds its way into the social and political arena.

Among the many uneasy alliances Dr. Katz draws our attention to, one that especially stands out is the conflict between the public's right to know the most up to date scientific information and the agendas of interest groups that may have strong pressure on administration policy. Especially compelling among Dr. Katz's examples is the strong impact of the second Bush administration's alliance with the religious right and its support of faith-based initiatives, each of which has apparently shaped dissemination and education policies regarding scientific research. In her essay on "Science Meets Fundamentalist Religion" (Chapter 10), Pat Shipman brings us back to the interface of science and religion in her reflections on the influence of religion on science education. Examples from the second Bush administration and its faith-based policies offer fresh and disconcerting evidence for Shipman's concerns that faith-based influences on science education can lead to profound detrimental effects on future scientific progress. Her reflections once again bring up concerns about accountability and authority – what are the implications for accountability when special groups such as fundamentalist religions (or, for that matter, political parties) choose to shape science education in our schools to their own needs and beliefs and how do members of both the scientific and media community play into these controversies?

Shipman builds her story around two primary examples – the 2005 incident in Dover, Pennsylvania and a similar 2006 incident in El Tejon, California – to limit the teaching of evolution, one of those scientific developments of the 19 century that Reverend Carr describes as changing the individual and societal "you" (along with Freud and Marx) in favor of Intelligent Design and creationism. Shipman extends her evidence to show how similar issues are arising world-wide in which various religious groups work to oppose the teaching of evolution in classrooms because it is not a proven scientific theory and contradicts basic tenets of creationism. It is important to note that the arguments in these many testing grounds are becoming more and more sophisticated. For example, the Biologic Institute in Redmond, Washington brings empirical methods and standard lab science to study intelligent design with the aim of developing a portfolio of research publications that may be viewed by legal authorities as a seemingly credible counter-voice to work based on theories of evolution.[13]

Like Smith, Klink, and Landwirth in Chapter 6, Shipman expands the perspectives on responsibility by placing accountability in the hands of both scientists and the media. In her examples of the growing debate between the "evolution" and "creationists" camps, she argues that the media has the responsibility to recognize the complexity of science, what she calls "the push and pull of scientific inquiry and the testing process." Driven in part by the need to capture the public's attention, reporters may emphasize apparently new, unexpected discoveries that have not yet stood up to the test of replicability, or they may emphasize conflicts between theories in an effort to play up the drama of a scientific issue. A hypothesis that does not yield positive findings or is disproven does not sell stories no matter how elegant the design or the creativity of the investigator. Shipman suggests that by adopting this more superficial attitude of going for the controversy or the dazzling finding, the science journalist abrogates his or her responsibility to tell the more complex – and truthful – story of scientific investigation.

Showing that science is incremental and that negative findings, rather than being reflective of errors or mistakes, are critical to narrowing and refining a question is key to educating the public about the nature of scientific knowledge and discovery. Further, publicizing how scientific knowledge is acquired is key to supporting educators in their moral obligation to teach their students that truth in science is not absolute and that the mark of a flexible and educated mind is the ability to embrace the complexity of how science evolves. At the same time, even as she insists that journalists have a moral obligation to change their standards for the "good story" as Shipman calls it, she also insists that scientists have an equally strong responsibility to upgrade their own skills for communicating with the media and the public. Scientists are not above the moral imperative to educate the public and to disseminate their work. Indeed, their accountability rests in sufficiently educating their communities so that their work may have serious debate and consideration instead of polarized, politicized debates. To assume this responsibility, scientists themselves need to become better communicators with the media and not relinquish their obligation to disseminate knowledge to professional writers. Full accountability requires a working partnership between scientists and writers, each embracing the need to show science in its complexity and knowledge in its dynamic evolution.

5 Getting the word out. Writing about research

Notions of moral responsibility and accountability bring us to questions of the ethical conduct of science and what are the pressures that may shape or distort scientific communication either to other scientists or to the public. Following closely on Wesley Carr's concerns in Chapter 9, Robert Levine makes clear, in very simple language, the ethical boundaries for the scientist and science writer: "to present an accurate account of the fruits of their efforts whether these presentations are to their colleagues or to the larger general public" (p. 148 in Chapter 12). This ethical charge includes making reasonable efforts to convey limitations of the work as well as the context in which the work is most accurately understood. For example, very often, findings from animal models are extrapolated to implications for humans without sufficient attention to the different contexts of such studies. The impact of a psychotropic drug on hippocampal neurons in the rat may be very different in the human and even more so in the adult man or woman suffering from chronic post-traumatic stress disorder with depression.

Levine builds on the "case history" method in the beginning sections of the book by offering his own series of historical vignettes in which the basic ethical tenets for publishing and disseminating scientific findings are either violated or shown to be very complex challenges to adhere to. He divides his examples along challenges common to the themes of many other authors in this volume – the rush to publish on the part of scientists, faulty interpretation of scientists by writers and reporters, deliberate disinformation disseminated by either scientists or writers, falsely exaggerated importance of a finding because of the publicity of a lawsuit or other associated events, differing interpretations of findings by experts from both science and the media resulting in conflicting, sometimes confusing accounts. Levine suggests that miscommunications between scientists and the media have many roots and that fortunately there are few examples of singularly nefarious intentions to mislead. He also reminds us that it is much easier to focus on what he calls the "external goods" – those factors more easily observable such as conflicts of interest, data fabrication, plagiarism – all those events that are reportable as clear ethical concerns or violations in science reporting. It is far harder to capture the "internal goods," those factors that describe the essence of the responsible scientist, and Levine argues that perhaps there will always be a divide between the practitioner and non-practitioner in understanding the essence of the good scientist. Such a divide makes it very difficult, if not impossible, for writers to report both accurately and comprehensively on the activities of a scientist.

If Levine's claim regarding the inherent chasm between the scientist and non-scientist is true, this poses ever greater challenges for the science writer. Laura Spinney continues themes, also highlighted by Smith, Klink, and Landwirth in Chapter 6, in her discussion of the many forces that may lead to a one-sided or at least slightly slanted account of a scientist's work, not the least of which is the need for science writers to "work" for their readership in creating an engaging story. It is not just popular magazines or newspapers, or more currently blog sites, that are impacted by these competing forces. Nor is the primary conflict always between the need for economic stability of the media source versus scientific accuracy. For example, just recently, the premier biomedical research publication, the *New England Journal of*

Medicine (NEJM), intentionally rushed the publication of a report on its web site that contained a flawed analysis of safety concerns for the diabetes drug, Avandia. This NEJM publication was deliberately timed to precede and overshadow a more careful evaluation of these same issues by the FDA. In beating the FDA in making this report public, the NEJM seemed to be continuing along the lines that it has in the past, in attempting to characterize the FDA as "impotent" and to "argue for legislation winding through congress that would increase regulatory hurdles for drug approvals."[14] This entire episode has been well described by Dr. Scott Gottlieb in a recent editorial published in the *Wall Street Journal*, and it underscores the role of the journals and journal editors as partners in protecting the veracity of scientific reports, or as accomplices in the rushed or flawed publication of new research in hopes of some secondary gain (e.g., notoriety for the journal, political clout in pushing for reforms).[15]

In Chapter 13, Spinney also underscores the responsibility of writers and their editors to ensure the veracity of their reporting even if it means they must delay the publication of a story. She illustrates the cascade of ill effects that can follow inaccurate or rushed science reporting in a number of examples including the purported impact of the three in one immunization for measles, mumps, and rubella (MMR) on increased risk for autism, a storyline generated through an interview with the primary author of a publication in *Lancet*. The human-interest or fear-factor was considerable in such a story, and thus many of the media sources picking up the account were written by non-science writers with little expertise in the area. The fire storm surrounding the initial story was sufficiently intense as to still leave question in the minds of many parents about whether or not to immunize their children. Spinney suggests that editors have a responsibility to seek writers with the most expertise to cover complex scientific issues, and that writers have an ethical responsibility to balance their accounts with statements of relative risk. For example, in the MMR controversy, little was said in most published accounts about the relative risk of the purported association compared to the risks of complications from any of the three illnesses covered by the vaccine. Spinney reminds us that science writers often have the difficult task of discerning not only the importance of a finding but also the accountability of the scientist whom they interview. Scientists are also subject to competing needs as discussed by a number of authors in this volume and motivated by the perceived benefits of notoriety may overstate the implications of their findings in ways that are sure to capture the interests of a writer needing a compelling story. For example, in 2002 in *Nature*, Catherine Verfaillie and her research team at the University of Minnesota described stem cells from mouse bone marrow as versatile as embryonic stem cells.[16] The finding ran counter to the accepted observation that adult stem cells can generally form only a narrow range of tissue types. Following other investigators' difficulties in replicating the work, reviewers from the *New Scientist* questioned plots describing the features of cell surfaces and stimulated an inquiry by the university.[17] The *New Scientist* 2006 review showed that some of the images in the *Nature* paper appeared in a second paper published at the same time but these images related to a different experiment. Verfaillie wrote to *Nature* cautioning that presentation of her team's results was flawed and the journal issue a correction through stressing that the original conclusions remained important. But in 2007, further inquiry by writers in the *New Scientist* reviewed a paper in *Blood* by the same research team[18] in which images were the same as those used in a US Patent application to support different findings. In this example, science writers took the responsibility to carefully review the evidence and to raise appropriate questions regarding the presentation of the findings. It is not that the media should assume a role of "policing" science reporting but rather that assuming the responsibility for accurate reporting also entails critically evaluating scientific reports, especially those that may appear so radically different from other reports or that fail the essential test of replicability. Indeed, such responsibility also belongs to editors of medical journals. A code of conduct aimed to identify bad practice and misconduct proposed by the Committee on Publication Ethics would enable editors to critically evaluate submitted manuscripts for these code violations.[19]

More attention to the accuracy of stories and to fairly representing relative risks when purported serious complications are linked with common biomedical practices should also be tempered with an understanding of the relative importance of a story. Spinney cautions us that sometimes a story makes front page news not because of its importance compared to another story from the same writer, but rather because the story size makes for better copy or the story itself presents a less complicated message. Some of the most recent and well publicized controversies such creationism versus evolutionary accounts may well have come to the forefront because of overly generous media coverage that pushed to the side other compelling scientific issues, a point similar to the concerns voiced by Shipman in Chapter 10. Spinney also agrees with her fellow science writer, Shipman, in her insistence that scientists also have an obligation, indeed a moral responsibility, to develop a working relationship with science writers so that they can educate writers and vice versa, writers can teach scientists how to effectively and responsibly communicate their work.

Many of the authors in this volume have called for changes in the practice of both scientists and reporters in disseminating the findings of research to both professional and general audiences. In his chapter, McConnell shows how changes in the publication of medical research itself will present additional challenges to the issues of authority and accountability. He also underscores the reality that nearly every finding from biomedical research no matter how flawed or questionable may find its way into print in some source. He introduces factors such as the prestige or impact factor of the journal into the discussion. Research gains significance in part because of the prestige of the journal (or other media source) in which it is published. There is also a feedback loop – publications in such prestigious journals are more often cited and these same journals gain further in prestige and visibility.

But there are inherent risks to an implicit ranking of a journal's impact especially for getting cutting edge research to broader audiences. As McConnell cites, over half of published biomedical research is not cited within 5 years of their publication. Thus, journals of high visibility cannot possibly represent even half of the emerging work in the biomedical sciences and hence, they may present a skewed view of the status of a topic or a field. Further, the cost of traditional publishing has increased considerably over the last decade and especially among higher impact journals. These escalating economic costs make it increasingly difficult for libraries and related research centers to afford access to a large, representative sample of biomedical journals.

A corollary to the concerns regarding how journals are selected by investigators as sources for publishing their work is how journals select submitted manuscripts for publication. McConnell outlines the growing concerns about the peer-review process, the heretofore perceived gold-standard for selection of scientific reports for publication. As he outlines, peer review is subject to many of the same competing interests and expectancies of human nature that we have discussed about reporting of science in the broader media. Competition and envy as well as conflicts of interests and constraints on time and expertise may seriously compromise the peer-review process especially in an error of rapid methodological and technological advances in the medical sciences. There is also the well-recognized trend for reviewers to recommend for publication articles with positive findings far more often than those with negative findings and/or articles from better recognized research groups.

Concerns about how journals and manuscripts are selected come together with the digital technology explosion to create new avenues for the publication of research that may well outpace traditional procedures such as measuring the number of times a manuscript is cited or blinded peer review. Web-based journals make it easier and more economical for researchers to gain access to a far greater range and number of publications. On-line publishing makes possible continuous peer review and commentary that is both transparent (i.e., not anonymous) and dynamic (i.e., in response to both the author and other reviewers). It also makes possible very rapid uploading of manuscripts and thus provides a solution to the often long wait between submission of manuscript and appearance in standard print. This means the newest findings are more rapidly available to other researchers and hence increases the rate of citation as well as increasing the possibility for productive and immediately relevant exchange of information with other laboratories. McConnell sees open access, open peer review, and open archiving as healthy, dynamic developments in the dissemination of biomedical research that will surely have positive effects as well on how science writers and reporters disseminate research findings. While it is not entirely clear to date whether or not open access publishing actually increases the number of citations for a research report[20] at least for biomedical sciences, it does seem clearer that open access does both speed up the rate of exchange among scientists and may offer new (and productive) opportunities for new collaborations.[21] How more rapid, more open publication will impact the media's reporting of scientific findings is not yet clear though if access to peer reviewers' comments is equally open, the opportunity for writers both to learn and to critically evaluate the findings may be enhanced considerably.

6 Final remarks

In a breathtakingly poignant and well-written essay, entitled "Conduct, Misconduct and the Structure of Science" Professors James Woodward and David Goodstein teach us that the most admirable of ethical principles – the very ideology that Western scientists are reared on – lie in direct conflict with the actual practice of science as an occupation and social endeavor.[22] These ethical tenets arise from the earliest theory of scientific method, as espoused by the English Renaissance philosopher and statesman, Sir Francis Bacon (1561–1626). Bacon, who was influenced in his thinking by Plato, asserted that scientists must act as disinterested observers of nature, with minds that are free of prejudices or preconceptions. With this belief held as central to competent practice as a scientist, Woodward and Goodstein offer a list of 15 corollary principles that conform to the Baconian ideal. As just a few examples, who among us would not agree with the following principles to govern ethical conduct as a scientist?

- A scientist should never be motivated to do science for personal gain, advancement, or other rewards.
- When an experiment or an observation gives a result contrary to the prediction of a certain theory, all ethical scientists must abandon that theory.
- Scientists must report what they have done so fully that any other scientist can reproduce the experiment or calculation. Science must be an open book, not an acquired skill.
- Financial support for doing science and access to scientific facilities should be shared democratically, not concentrated in the hands of a favored few.

Woodward and Goodstein argue that although we ideally aspire to live by these rules, and the 11 others like them that they list in their article, these principles are actually "defective" and they undermine the logical structure of science.

Scientists, journalists, editors, and others engaged in the reporting of new discoveries are all persons that are motivated by the opportunity to make a lasting mark on their communities, to be remembered by their peers for good work, to provide well for their children and families, and achieve other markers of success as a professional, spouse, and care-provider. Although such expected desires often lead to temptations to publish too soon, to stretch the truth, to over-simplify a problem or solution, as well as more patently unethical subterfuge of the scientific process (e.g., stealing data, intentional misrepresentation), these same motivating factors work to propel scientific discoveries forward. Woodward and Goodstein go so far as to suggest that "behavior that may seem at first glance morally unattractive, such as the aggressive pursuit of economic self-interest, can, in a properly functioning system, produce results that are generally beneficial." This is a premise supported by some authors in this volume, such as Dr. Doogan in Chapter 8, and viewed with caution and concern by others, such as Prof. Kassirer in Chapter 7.

There are good reasons to believe that the Woodward and Goodstein argument is a valid one. As an example, the philosopher, Philip Kitcher has made the point that because the first person who makes a scientific discovery usually attracts most of the credit, this encourages investigators to pursue a broader range of different lines of inquiry – including avenues of research that may be thought by many to be associated with smaller chances of eventual success – so that others maximize their own chances of meeting with the same success with an "unpolished gem" that they might uncover.[23] This, in turn, leads to a wider diversity of concurrent lines of scientific inquiry, and as a result, it is more likely that an incorrect majority opinion in a given field might be overturned more rapidly. Each of the logical tenets listed in the Woodward and Goodstein article, based on Baconian philosophy of scientific inductivism, can be proven to be fraught with problems, and in doing so, they call into question our core ethical ideals. They argue that "in designing institutions and regulations to discourage scientific misconduct, that we not introduce changes that disrupt the beneficial effects that competition and a concern for credit and reputation bring with them."[21]

It is certainly true that an important motive in a number of recent cases of data fabrication or misrepresentation, such as the Verfaillie incident cited above, has been a desire on the part of one or more persons to establish priority and to receive credit for a discovery. No doubt that a great amount of fraud in the conduct and reporting of science can be traced directly to such motives, and to the highly competitive nature of practice as a scientist. The question for all of us is a simple one. How do we reach our aspirations to succeed, and to protect the benefits of healthy competition and reward, and at the same time protect the veracity of our work, the trust of our communities, and the social benefits of our discoveries when we bring our information from the lab bench to the daily newspaper? The essays in this volume may raise more questions than they answer – but these questions are all central to our professional lives, our conduct, and ultimately they are crucial issues to consider for any individual who is fortunate enough to be in a position to make, or report on, important discoveries.

References and Notes

1 Introduction. The Ethics of Scientific Disclosure

1 Cho, K.Y., Wirth, D.P., and Lobo, R.A. (2001). Does prayer influence the success of *in vitro* fertilization-embryo transfer? *J. Reprod. Med.* 46, 781–787.

2 Schorr, M. (October, 2001). Prayer may boost *in-vitro* success, study suggests. *Reuters News Service* .

3 Eisner, R. Prayer may influence *in vitro* fertilization success. *Columbia News*. This document remained on the Public Affairs News page of Columbia University Internet site for more than two years after the publication of the Cha/Wirth/Lobo study (www.columbia.edu/cu/news).

4 Nagourney, E. (02 October, 2001). Study links prayer and pregnancy. *New York Times* .

5 Johnson, T. (02 October, 2001). Praying for pregnancy: Study says prayer helps women get pregnant. ABC television show. *New York Times* .

6 Can prayer heal? Scientists suggest recovery may be the hand of God at work, *ABC News*, 13 August, 2001, at: http://abcnews.go.com/

7 Flamm, B.L. (2002). Faith healing by prayer: Review of Cho, KY, Wirth, DP, Lobo, RA. Does prayer influence the success of *in vitro* fertilization-embryo transfer? *Sci. Review Alt. Med.* 6(1), 47–50.

8 Flamm, B.L. (2004). Faith healing confronts modern medicine. *Sci. Review Alt. Med.* 8(1), 9–14.

9 Flamm, B.L. (September–October, 2004). The Columbia University "miracle" study: flawed and fraud. *Skeptical Inquirer Magazine*, at: www.looksmarttrends.com/p/articles/mi_m2843/is_5_28/ai_n6194238

10 Harris, D. (May 30, 2004). Exposed: Conman's role in prayer-power IVF "miracle". *The [London] Observer*.

11 Carmone, M.A. (December, 2001). Letter to Thomas Q. Morris, MD, Vice President for Health Sciences Division, regarding possible noncompliance with DHHS regulations for protection of human subjects in the conduct of the Cha *et al.* study, at: http://ohrp.osophs.dhhs.gov/detrm_letrs/dec01f.pdf

12 Diefenbach, G.J. *et al.* (1999). Portrayal of lobotomy in the popular press: 1935–1960. *J. Hist. Neurosci.* 8, 60–69.

13 Garrett, J.M. and Bird, S.J. (2000). Ethical issues in communicating science. *Sci. Eng. Ethics* 6, 435–442.

14 Racine, E., van der Loos, H.Z.A., and Illes, J. (2007). Internet marketing of neuroproducts: New practices and healthcare policy challenges. *Camb. Q. Healthc. Ethics* 16, 181–194.
15 Illes, J. and Bird, S. (2006). Neuroethics: A modern context for ethics in neuroscience. *Trends Neurosci.* 29, 511–517.
16 Racine, E., Bar-Ilan, O., and Illes, J. (2005). fMRI in the public eye. *Nat. Rev. Neurosci.* 6(2), 159–164.

2 Medicine, Media, and the Dramaturgy of Biomedical Research

1 Geison, G.L. (1995). *The Private Science of Louis Pasteur* pp. 181, 220. Princeton University Press, Princeton.
2 Karpf, A. (1988). *Doctoring the Media: The Reporting of Health and Medicine.* Routledge, New York.
3 LaFollette, M.D. (1990). *Making Science Our Own: Public Images of Science 1910–1955.* University of Chicago Press, Chicago.
4 Nelkin, D. (1995). *Selling Science: How the Press Covers Science and Technology.* W.H. Freeman, New York.
5 Osler, W. (1904). *Aequanimitas* p. 144. Blakiston Company, Philadelphia.
6 Cathell, D.W. (1916). *Book on the Physician Himself* p. 144. F.A. Davis Company, Philadelphia.
7 Tomes, N. (1998). *The Gospel of Germs: Men, Women, and the Microbe in American Life* p. 123. Harvard University Press, Cambridge.
8 Mendelsohn, J.A. (1996). Cultures of bacteriology: Formation and transformation of a science in France and Germany, 1870–1914. Ph.D. Dissertation, Princeton University, pp. 291, 307.
9 Warner, M. (1985). Hunting the yellow fever germ: The principle and practice of etiological proof in late nineteenth-century America. *Bull. Hist. Med.* 59, 361–382.
10 Hansen, B. (1998). America's first medical breakthrough: How popular excitement about a French rabies cure in 1885 raised new expectations for medical progress. *Am. Hist. Rev.* 103, 373–418.
11 Hansen, B. (1999). New images of a new medicine: Visual evidence for the widespread popularity of therapeutic discoveries in America after 1885. *Bull. Hist. Med.* 73, 629–678.
12 Weindling, P. (1989). *Health, Race and German Politics between National Unification and Nazism, 1870–1945* p. 650. Cambridge University Press, Cambridge.
13 Warner, J.H. (1995). The history of science and the sciences of medicine. *Osiris* 10, 164–193.
14 Rogers, N. (2005). Vegetables on parade: American medicine and the child health movement in the jazz age. In *Children's Health Issues in Historical Perspective* (C. Krasnick and V. Strong-Boag, eds.). Wilfred Laurier University Press, Waterloo, Ontario, pp. 23–71.
15 Pernick, M.S. (1978). Thomas Edison's tuberculosis films: Mass media and public propaganda. *Hastings Center Report* 8, 21–27.
16 Tomes, N. (2001). Merchants of health: Medicine and consumer culture in the United States. *J. Am. Hist.* 88, 23–71.

17 Toon, E. (1998). Managing the conduct of the individual life: Public health education and American public health, 1910–1940. Ph.D. Dissertation, University of Pennsylvania.
18 Burnham, J.C. (1987). *How Superstition Won and Science Lost: Popularizing Science and Health in the United States* pp. 76, 196, 199, 235. Rutgers University Press, New Brunswick.
19 Rhees, D.J. (1979). A new voice for science: Science service under Edwin E. Slosson, M.A. Thesis, University of North Carolina, pp. 1921–1929, at: http://scienceservice.si.edu/thesis/
20 Summers, W.C. (1991). On the origins of the science in Arrowsmith: Paul de Kruif, Felix d'Herelle, and phage. *J. Hist. Med. Allied Sci.* 46, 315–332.
21 Duffus, R.L. (February 26, 1926). Adventurers with test tube and microscope. *New York Times Book Review*, p. 10.
22 Chernin, E. (1988). Paul de Kruif's microbe hunters and an outraged Ronald Ross. *Rev. Infect. Dis.* 10, 661–667.
23 Rosenberg, C.E. (1976). Martin Arrowsmith: The scientist as hero. *No Other Gods: On Science and American Social Thought*. Johns Hopkins University Press, Baltimore, pp. 123–131.
24 Summers, W.C. (1998). Microbe hunters revisited. *Int. Microbiol.* 1, 65–68.
25 LeBlanc, T.J. (1925). Review of Arrowsmith. *Science* 61, 632–634.
26 American Film Institute. (1998). Arrowsmith. *The American Film Institute Catalog of Motion Pictures*. University of California Press, Berkeley, pp. 85–86.
27 Lederer, S.E. and Rogers, N. (2000). Media. In *Medicine in the Twentieth Century* (R. Cooter and J. Pickstone, eds.). Harwood Academic Press, Amsterdam, pp. 487–502.
28 Lederer, S.E. and Parascandola, J. (1998). Screening syphilis: Dr. Ehrlich's magic bullet meets the public health service. *J. Hist. Med. Allied Sci.* 53, 345–370.
29 Rogers, N. (2007). American medicine and the politics of film making: Sister Kenny (RKO, 1946). In Medicine's Moving Pictures: Medicine, Health, and Bodies in American Film and Television (L.J. Reagan, N. Tomes, and P.A. Treichler, eds.), University of Rochester Press, Rochester, pp. 199–238.
30 Rogers, N. (2004). Sister Kenny goes to Washington: Polio, populism, and medical politics in postwar America. In *The Politics of Healing: Histories of Alternative Medicine in Twentieth-Century America* (R.D. Johnston, ed.). Routledge Press, New York, pp. 97–116.
31 Shryock, R.H. (1947). *American Medical Research: Past and Present*. Commonwealth Fund, New York.
32 Apple, R.D. and Apple, M.W. (1993). Screening science. *Isis* 84, 750–754.
33 Brandt, A.M. and Gardner, M. (2000). The golden age of medicine. In *Medicine in the Twentieth Century* (R. Cooter and J. Pickstone, eds.). Harwood Academic Press, Amsterdam, pp. 21–32.
34 Burnham, J.C. (1982). American medicine's golden age: What happened to it? *Science* 215, 1474–1479.
35 McLeod, K. (2007). Health matters: Public understandings of health in 1950s America. Ph.D. Dissertation, New Haven, Yale University.
36 Hansen, B. (2004). Medical history for the masses: How American comic books celebrated heroes of medicine in the 1940s. *Bull. Hist. Med.* 78, 148–191.
37 Duffin, J. and Li, A. (1995). Great moments: Parke, Davis and company and the creation of medical art. *Isis* 86, 1–19.

38 Turow, J. (1989). *Playing Doctor: Television, Storytelling, and Medical Power*. Oxford University Press, New York.

39 Brandt, A.M. (2001). Polio, politics, publicity, and duplicity: The salk vaccine and the protection of the public. In *Major Problems in the History of American Medicine and Public Health* (J.H. Warner and J.M. Tighe, eds.). Houghton Mifflin Company, Boston, pp. 451–457.

3 Delgado's Brave Bulls

1 Lashley, K.S. (1929). *Brain mechanisms and intelligence: A quantitative study of injuries to the brain*. University of Chicago Press, Chicago.

2 Ewald, J.R. (1898). Ueber künstlich erzeugte Epilepsie. *Berliner Klin. Wochenschr* vol. 35, p. 689.

3 Hess, W.R., 1932 German publication, cited in Delgado, J.M.R. (1965). Evolution of physical control of the brain. The James Arthur Lecture on the Evolution of the Human Brain. American Museum of Natural History, New York, p. 6.

4 Gibbs, E.L. and Gibbs, F.A. (1936). A purring center in the cat's brain. *J. Comp. Neurol.* 64, 209–211.

5 Valenstein, E.S. (1973). *Brain Control. A Critical Examination of Brain Stimulation and Psychosurgery* p. 389. . John Wiley & Sons, New York.

6 Olds, J. and Milner, P. (1954). Positive reinforcement produced by electrical stimulation of the septal area and other regions of the rat brain. *J. Comp. Physiol. Psychol.* 47, 419–427.

7 Baumeister, A.A. (2006). Serendipity and the cerebral localization of pleasure. *J. History Neurosci.* 15, 92–98.

8 Valenstein, E.S. (1973). *Brain Control. A Critical Examination of Brain Stimulation and Psychosurgery* pp. 34–40, 46, 54–56, 61–62. John Wiley & Sons, New York.

9 Heath, R.G. (1954). *Studies in Schizophrenia. A Multidisciplinary Approachs to Mind-Brain Relationships*. Harvard University Press, Cambridge.

10 Tierney, A.J. (2000). Egas Moniz and the origins of psychosurgery: A review commemorating the 50th anniversary of Moniz's Nobel Prize. *J. History Neurosci.* 9, 22–36.

11 O'Neill, J.J. (1953). Overeat? Blame the hypothalamus. New York Herald Tribune, Section 2, May 10, p. 5.

12 Yale University News Bureau, New Haven, CT (1966). Press Release #171(P). December 01.

13 Horgan, J. (2005). The forgotten era of brain chips. *Scientific American* 293(4), 66–73. October.

14 Delgado, J.M.R., Hamlin, H., and Chapman, W.P. (1952). Technique of intracranial electrode implacement for recording and stimulation and its possible therapeutic value in psychotic patients. *Conf. Neurol.* 12, 315–319.

15 Delgado, J.M.R., Mark, V., Sweet, W., Ervin, F., Weiss, G., Bach-y-Rita, G., and Hagiwara, R. (1968). Intracerebral radio stimulation and recording in completely free patients. *J. Ner. Mental Dis.* 147, 329–340.

16 Delgado, J.M.R., Obrador, S., and Martin-Rodriguez, J.G. (1973). Two-way radio communication with the brain in psycho-surgical patients. In *Surgical approaches in Psychiatry* (L.V. Laitinen and K.E. Livingston, eds.). Medical & Technical Publishing, Lancaster, England.

17 Yale University News Bureau, New Haven, CT (1953). Press Release #535. April 09.
18 New technique is reported in study of brain. Yale scientist seeks more effective way to treat mental disease. *New Haven Evening Register*, April 09, 1953.
19 [Attribution: "W.K."] Electric technique for mental cases. *The New York Times*, Sect. E, April 12, 1953, p. 11, 12.
20 Ocean of the Mind, April 13, 1953. Time Magazine Archives: http://www.time.com/time/magazine/article/0,9171,818225-1,00.html
21 Delgado, J.M.R. (1965). *Evolution of physical control of the brain. The James Arthur Lecture on the Evolution of the Human Brain*. American Museum of Natural History, New York.
22 Delgado, J.M.R. (1969). *Physical Control of the Mind. Toward a Psychocivilized Society* (World Perspectives Series, Vol. 41). Harper & Row, New York.
23 Baumeister, A.A. (2000). The Tulane electrical brain stimulation program. A historical case study in medical ethics. *J. Hist. Neurosci.* 9, 262–278.
24 Osmundsen, J.A. (1965). "matador" with a radio stops wired bull. The New York Times, May 17, pp. 1 & 20.
25 Eibl-Eibesfeldt, I. (1967). Ontogenetic and maturational studies of aggressive behavior. In *Aggression and Defense. Neural Mechanisms and Social Patterns (Brain Function, Volume V). UCLA Forum in Medical Sciences, Number 7* (C.D. Clemente and D.B. Lindsley, eds.). University of California Press, Los Angeles, pp. 88–91.
26 Delgado, J.M.R. (1967). Aggression and defense under cerebral radio control. In *Aggression and Defense. Neural Mechanisms and Social Patterns (Brain Function, Volume V). UCLA Forum in Medical Sciences, Number 7* (C.D. Clemente and D.B. Lindsley, eds.). University of California Press, Los Angeles, pp. 171–193.
27 Forman, D. and Ward, J. (1957). Responses to electrical stimulation of the caudate nucleus in cats in chronic experiments. *J. Neurophysiol.* 20, 230–244.
28 McLennan, H., Emmons, P.R., and Plummer, P.M. (1964). Some behavioral effects of stimulation of the caudate nucleus in unrestrained cats. *Canadian J. Physiol. Pharmacol.* 42, 329–339.
29 Delgado, J.M.R. (1964). Toros radiodirigidos. *Gaceta Ilustrada*, No. 393, pp. 84–89.
30 Delgado, J.M.R., Castejon, F.J., and Santisteban, F. (1964). *Radioestimulatíon cerebral en toros de lidia. VIII Reun. Nacional Soc.* Ciencias Fisiológicas, Madrid, Febrero.
31 Waggoner, K. (1970). Psychocivilization or elictroligarchy: Dr. Delgado's amazing world of ESB. *Yale Alumni Magazine*, 33(4), 20–25. January.
32 Snyder, D.R. (2005). *Personal Communication* (e-mail), July 19.
33 Thompson, R.A. and Nelson, C.A. (2001). Developmental science and the media. *Am. Psychol.* 56, 5–15.

4 Adrenal Transplantation for Parkinson's Disease

1 US Congress, Office of Technology Assessment. (1990). *Neural Grafting: Repairing the Brain and Spinal Cord* OTA-BA-462. September. US Government Printing Office, Washington, DC.
2 Parkinson, J. (1817). *An Essay on the Shaking Palsy*.
3 Dunnett, S.B., Bjorklund, A., Schmidt, R.H. *et al.* (1983). Intracerebral grafting of neuronal cell suspensions, IV: Behavioral recovery in rats with unilateral

6-OHDA lesions following implantation of nigral cell suspensions in different forebrain sites. *Acta Physiol. Scandinavia* 522(supp), 29–37.

4 Dunnett, S.B., Hernandez, T.D., Summerfield, A. *et al.* (1988). Grant-derived recovery from 6-OHDA lesions: specificity of ventral mesencephalic graft tissues. *Exp. Brain Res.* 71, 411–424.

5 Redmond, D.E., Sladek, J.R. Jr., Roth, R.H. *et al.* (1986). Fetal neuronal grafts in monkeys given methylphrnyl-tetrahydropyridine. *Lancet* 1, 1125–1127.

6 Freed, W.J., Morihisa, J.M., Spoor, E. *et al.* (1981). Transplanted adrenal chromaffin cells in rat brain reduce lesion-induced rotational behavior. *Nature* 292, 351–352.

7 Backlund, E.O., Granberg, P.O., Hamberger, B. *et al.* (1985). Transplantation of adrenal medullary tissue to striatum in Parkinsonism, First clinical trials. *J. Neurosurg.* 62, 169–173.

8 Madrazo, I., Drucker-Colin, R., Diaz, V. *et al.* (1987). Open microsurgical autograft of adrenal medulla to the right caudate nucleus in two patients with intractable Parkinson's disease. *New Eng. J. Med.* 316, 831–834.

9 Sullivan, W. (2007). Transplant brings dramatic gains for 2 Parkinson's victims. *The New York Times, Health Section* 21. January.

10 Associated Press. (2007). Transplant is first in nation. *The New York Times, Health Section* 21. January 21.

11 Madrazo, I., Franco-Bourland, R., Aguilera, M. *et al.* (1991). Autologous adrenal medullary, fetal mesencephalic, and fetal adrenal brain transplantation in Parkinson's disease: A long-term postoperative follow-up. *J. Neural Transp. Plas.* 2(3(4)), 157–164.

12 Waxman, M.J., Morantz, R.A., Koller, W.C., Paone, D.B., and Nelson, P.W. (1991). High incidence of cardiopulmonary complications associated with implantation of adrenal medullary tissue into the caudate nucleus in patients with advanced neurologic disease. *Crit. Care Med.* 19(2), 181–186.

13 Schvarcz, J.R., Devoto, M., Meiss, R., Torrieri, A., Genero, M., and Armando, I. (1990). Multiloci stereotactic transplantation of autologous adrenal medullary tissue to the putamen and caudatum in Parkinson's disease. *Proceeding of the Xth Meeting of the World Society for Stereotactic and Functional Neurosurgery*, Maebashi, Japan, October 1989. *Stereot. Funct. Neuros.* 54, 55, 277–281.

14 Pezzoli, G., Motti, E., Zecchinelli, A., Ferrante, C. *et al.* (1990). Adrenal medulla autograft in 3 Parkinsonian patients: Results using two different approaches (Chapter 76). In *Progress in Brain Research*, Vol. 82 (V. Dunnett and S.J. Richards, eds.). Elsevier B.V., Amsterdam, pp. 677–682.

15 Berry, P. and Ward-Smith, P.A. (1988). Adrenal medullary transplant as a treatment for Parkinson's disease: Perioperative considerations. *J. Neurosci. Nurs.* 20(6), 356–361.

16 Velasco, F., Velasco, M., Rodriguez Cuevas, H., Jurado, J., Olvera, J., and Jimenez, F. (1991). Autologous adrenal medullary transplants in advanced Parkinson's disease with particular attention to the selective improvement in symptoms. *Stereot. Funct. Neuros.* 57, 195–212.

17 Gildenberg, P.L., Pettigrew, L.C., Merrell, R., Butler, I., Conklin, R., Katz, J., and de France, J. (1990). Transplantation of adrenal medullary tissue to caudate nucleus using stereotactic techniques. *Stereot. Funct. Neuros.* 54 & 55, 268–271.

18 Goetz, C.G., Stebbins, G.T., Klawans, H.L. and the United Parkinson Foundation Neural Transplantation Registry. *et al.* (1991). United Parkinson Foundation

Nurotransplantation Registry on Adrenal Medullary Transplants: Presurgical, and 1- and 2-year follow-up. *Neurology* 41, 1719–1722.

19 Richter, L. (1988). Doctor in Mexico defends his innovative transplant procedures. *The New York Times, Health Section* 30. August.

20 Goetz, C.G., Olanow, C.W., Koller, W.C. *et al.* (1989). Multicenter study of autologous adrenal medullary transplantation to the corpus striatum in patients with advanced Parkinson's disease. *New Eng. J. Med.* 320(6), 337.

21 Diamond, S.G., Markham, C.H., Rand, R.W., Becker, D.P., and Treciokas, L.J. (1994). Four-year follow-up of adrenal-to-brain transplants in Parkinson's disease. *Arch. Neurol.* 51, 559–563.

22 Takeuchi, J., Takebe, Y., Sakakura, T. *et al.* (1990). Adrenal medulla transplantation into the putamen in Parkinson's disease. *Neurosurgery* 26(3), 499–503.

23 Bakay, R.A.E., Watts, R.L., Freeman, P.M. *et al.* (1990). Preliminary report on adrenal-brain transplantation for Parkinsonism in man. *Stereot. Funct. Neuros.* 54 & 55, 312–323.

24 Garcia-Flores, E., Decanini, H.L., Flores-Salazar, M. *et al.* (1990). Is autologous transplant of adrenal medulla into the striatum and effective therapy for Parkinson's disease? (Chapter 72). In *Progress in Brain Research,* Volume 82 (V. Dunnett and S.J. Richards, eds.). Elsevier B.V., Amsterdam, pp. 643–655.

25 Bakay-Roy, A.E. and the Analysis and Publications Subcommittee. (1990). Preliminary report on adrenal medullary grafting from the American Association of Neurological Surgeons Graft Project (Chapter 67). In *Progress in Brain Research,* Vol. 82 (V. Dunnett and S.J. Richards, eds.). Elsevier B.V., Amsterdam, pp. 603–610.

5 The Use of SSRls in the Treatment of Childhood Depression: A Scientific Dialectic

1 National Institute for Health and Clinical Excellence. (2005). *Depression in Children and Young People: Identification and Management in Primary, Community and Secondary Care.* National Institute for Health and Clinical Excellence, London.

2 Birmaher, B., Arbelaez, C., and Brent, D. (2002). Course and outcome of child and adolescent major depressive disorder. *Child Adolesc. Psychiatr. Clin. N. Am.* 11(3), 619–637.

3 Dunn, V. and Goodyer, I.M. (2006). Longitudinal investigation into childhood- and adolescence-onset depression: psychiatric outcome in early adulthood. *Br. J. Psychiat.* 188, 216–222.

4 American Psychiatric Association. (2006). *Diagnostic and Statistical Manual of Mental Disorders (DSM-IV),* 4th ed. American Psychiatric Association, Washington, DC.

5 Birmaher, B., Williamson, D.E., Dahl, R.E., Axelson, D.A., Kaufman, J., Dorn, L.D. *et al.* (2004). Clinical presentation and course of depression in youth: does onset in childhood differ from onset in adolescence? *J. Am. Acad. Child Adolesc. Psychiat.* 43(1), 63–70.

6 American Academy of Child and Adolescent Psychiatry. (in press). Practice parameters for the assessment and treatment of children and adolescents with depressive disorders. *J. Am. Acad. Child Adolesc. Psychiat.*

7 Jacobs, D. and Cohen, D. (1999). What is really known about psychological alterations produced by psychoactive drugs? *Int. J. Risk Saf. Med.* 12, 37–47.
8 Moncrieff, J. (2002). The antidepressant debate. *Br. J. Psychiatr.* 180, 193–194.
9 Cohen, D. and Jacobs, D. (2007). Randomized controlled trials of antidepressants: clinically and scientifically irrelevant. *Debates Neuorsci.* 1, 44–54.
10 Breggin, P.R. (1997). *Brain Disabling Treatments in Psychiatry*. Springer Publishing, New York.
11 Fineberg, N.A., Hawley, C.J., and Gale, T.M. (2006). Are placebo-controlled trials still important for obsessive compulsive disorder? *Prog. Neuropsychopharmacol. Biol. Psychiatr.* 30(3), 413–422.
12 Bridge, J.A., Iyengar, S., Salary, C.B., Barbe, R.P., Birmaher, B., Pincus, H.A. *et al.* (2007). Clinical response and risk for reported suicidal ideation and suicide attempts in pediatric antidepressant treatment: a meta-analysis of randomized controlled trials. *JAMA* 297(15), 1683–1696.
13 Smith, R. (2005). Medical journals are an extension of the marketing arm of pharmaceutical companies. *PLoS Med.* 2(5), e138.
14 Medaware, C. and Hardon, A. (2004). *Medicines Out of Control? Antidepressants and the Conspiracy of Goodwill*. Aksant Academic Publishers, London.
15 De Los Reyes, A. and Kazdin, A.E. (2005). Informant discrepancies in the assessment of childhood psychopathology: a critical review, theoretical framework, and recommendations for further study. *Psychol. Bull.* 131(4), 483–509.
16 De Los Reyes, A. and Kazdin, A.E. (2006). Conceptualizing changes in behavior in intervention research: the range of possible changes model. *Psychol. Rev.* 113(3), 554–583.
17 National Institute of Clinical Excellence. (2004). *Treatment and Management of Depression in Primary and Secondary Care*. National Institute of Clinical Excellence, London.
18 Roth, A. and Fonagy, P. (2005). *What works for whom? A Critical Review of Psychotherapy Research*, 2nd edition. Guilford Press, New York.
19 Caspi, A., Sugden, K., Moffitt, T.E., Taylor, A., Craig, I.W., Harrington, H. *et al.* (2003). Influence of life stress on depression: moderation by a polymorphism in the 5-HTT gene. *Science* 301(5631), 386–389.
20 Kessler, R.C., Berglund, P., Demler, O., Jin, R., Koretz, D., Merikangas, K.R. *et al.* (2003). The epidemiology of major depressive disorder: results from the National Comorbidity Survey Replication (NCS-R). *JAMA* 289(23), 3095–3105.
21 Mezzich, J.E., Kirmayer, L.J., Kleinman, A., Fabrega, H. Jr, Parron, D.L., Good, B.J. *et al.* (1999). The place of culture in DSM-IV. *J. Nerv. Ment. Dis.* 187(8), 457–464.
22 Kirsch, I., Moore, T., Scoboria, A., and Nicholls, S. (2002). The emperor's new drugs: An analysis of antidepressant medication data submitted to the US Food and Drug Administration. *Prevention and Treatment, 5*(article 23), at http://www.journals.apa.org/prevention/volume5/pre0050023a.html.
23 Jureidini, J.N., Tonkin, A.L., and Mansfield, P.R. (2004). TADS study raises concerns (letter). *Br. Med. J.* 329, 1343–1344.
24 Petovka, E., Quitkin, F.M., McGrath, P.J. *et al.* (2000). A method to quantify rater bias in antidepressant trials. *Neuropsychopharmacology* 22, 559–565.
25 Moncrieff, J., Wessely, S., and Hardy, R. (2004). Active placebos versus antidepressants for depression (Cochrane Review). *Cochrane Libr.* 1.

26 Desbiens, N.A. (2002). In randomized controlled trials, should subjects in both placebo and drug groups be expected to guess that they are taking the drug 50% of the time? *Med. Hypot.* 59(3), 227–232.

27 Rothwell, P.M. (2005). External validity of randomised controlled trials: to whom do the results of this trial apply? *Lancet* 365(9453), 82–93.

28 Hampton, J.R. (2002). Size isn't everything. *Stat. Med.* 21, 2807–2814.

29 Feinstein, A.R. and Horwitz, R.I. (1997). Problems in the "evidence" of "evidence based medicine". *Am. J. Med.* 103, 529–535.

30 Mosholder, A.D. and Willy, M. (2006). Suicidal adverse events in pediatric randomized, controlled clinical trials of antidepressant drugs are associated with active drug treatment: a meta-analysis. *J. Child. Adolesc. Psychopharmacol.* 16(1(2)), 25–32.

31 Weintrob, N., Cohen, D., Klipper-Aurbach, Y., Zadik, Z., and Dickerman, Z. (2002). Decreased growth during therapy with selective serotonin reuptake inhibitors. *Arch. Pediatr. Adolesc. Med.* 156(7), 696–701.

32 Garland, E.J. and Baerg, E.A. (2001). Amotivational syndrome associated with selective serotonin reuptake inhibitors in children and adolescents. *J. Child Adolesc. Psychopharmacol.* 11, 181–186.

33 Jain, B., Birmaher, M., and Garcia, M.al. (1992). Fluoxetine in children and adolescents with mood disorders: a chart review of efficacy and adverse reactions. *J. Child. Adolesc. Psychopharmacol.* 2, 259–265.

34 Wilens, T.E., Biederman, J., Kwon, A., Chase, R., Greenberg, L., Mick, E. *et al.* (2003). A systematic chart review of the nature of psychiatric adverse events in children and adolescents treated with selective serotonin reuptake inhibitors. *J. Child Adolesc. Psychopharmacol.* 13(2), 143–152.

35 Clavenna, A., Rossi, E., Derosa, M., and Bonati, M. (2007). Use of psychotropic medications in Italian children and adolescents. *Eur. J. Pediatr.* 166(4), 339–347.

36 Hammad, T.A., Laughren, T., and Racoosin, J. (2006). Suicidality in pediatric patients treated with antidepressant drugs. *Arch. Gen. Psychiat.* 63(3), 332–339.

37 US Food and Drug Administration. (2005). Relationship between psychotropic drugs and pediatric suicidality: review and evaluation of clinical data, at http://www.fda.gov/ohrms/dockets/ac/04/slides/4006S1_08_Posner_files/frame.htm

38 Whittington, C.J., Kendall, T., Fonagy, P., Cottrell, D., Cotgrove, A., and Boddington, E. (2004). Selective serotonin reuptake inhibitors in childhood depression: Systematic review of published versus unpublished data. *Lancet* 363(9418), 1341–1345.

39 Editorial. (2004). Depressing research. *Lancet* 363, 1335.

40 Giles, J. (2006). Drug trials: Stacking the deck. *Nature* 440, 270–272.

41 DeAngelis, C.D., Drazen, J.M., Frizelle, F.A., Haug, C., Hoey, J., Horton, R. *et al.* (2005). Is this clinical trial fully registered? A statement from the International Committee of Medical Journal Editors. *JAMA* 293(23), 2927–2929.

42 Freedman, R., Lewis, D.A., Michels, R., Pine, D.S., Schultz, S.K., Tamminga, C. *et al.* (2006). Conflict of interest, round 2. *Am. J. Psychiatr.* 163(9), 1481–1483.

43 Cosgrove, L., Krimsky, S., Vijayaraghavan, M., and Schneider, L. (2006). Financial ties between DSM-IV panel members and the pharmaceutical industry. *Psychother. Psychosom.* 75(3), 154–160.

44 Fontanarosa, P.B., Rennie, D., and DeAngelis, C.D. (2004). Postmarketing surveillance – lack of vigilance, lack of trust. *JAMA* 292(21), 2647–2650.

45 Schulman, K.A., Seils, D.M., Timbie, J.W., Sugarman, J., Dame, L.A., Weinfurt, K.P. *et al.* (2002). A national survey of provisions in clinical-trial agreements between medical schools and industry sponsors. *N. Engl. J. Med.* 347(17), 1335–1341.

46 Mirowski, P. and Van Horn, R. (2005). The contract research organization and the commercialization of scientific research. *Soc. Stud. Sci.* 35(4), 503–548.

47 DeAngelis, C.D. (2006). The influence of money on medical science. *JAMA* 296(8), 996–998.

48 Melander, H., Ahlqvist-Rastad, J., Meijer, G., and Beermann, B. (2003). Evidence b(i)ased medicine – selective reporting from studies sponsored by pharmaceutical industry: Review of studies in new drug applications. *Br. Med. J.* 326(7400), 1171–1173.

49 Bennett, C.L., Nebeker, J.R., Lyons, E.A., Samore, M.H., Feldman, M.D., McKoy, J.M. *et al.* (2005). The research on adverse drug events and reports (RADAR) project. *JAMA* 293(17), 2131–2140.

50 Glass, R.M. (2005). Fluoxetine, cognitive-behavioral therapy, and their combination for adolescents with depression: Treatment for adolescents with depression study (TADS) randomized controlled trial. *J. Pediatr.* 146(1), 145.

51 Simon, G.E., Savarino, J., Operskalski, B., and Wang, P.S. (2006). Suicide risk during antidepressant treatment. *Am. J. Psychiatr.* 163(1), 41–47.

52 Valuck, R.J., Libby, A.M., Sills, M.R., Giese, A.A., and Allen, R.R. (2004). Antidepressant treatment and risk of suicide attempt by adolescents with major depressive disorder: A propensity-adjusted retrospective cohort study. *CNS Drugs* 18(15), 1119–1132.

53 Cheung, A.H., Emslie, G.J., and Mayes, T.L. (2005). Review of the efficacy and safety of antidepressants in youth depression. *J. Child Psychol. Psychiatr.* 46(7), 735–754.

54 Martin, A., Young, C., Leckman, J.F., Mukonoweshuro, C., Rosenheck, R., and Leslie, D. (2004). Age effects on antidepressant-induced manic conversion. *Arch. Pediatr. Adolesc. Med.* 158(8), 773–780.

55 Boylan, K., Romero, S., and Birmaher, B. (2007). Psychopharmacologic treatment of pediatric major depressive disorder. *Psychopharmacology (Berl)* 191(1), 27–38.

56 March, J., Silva, S., Petrycki, S., Curry, J., Wells, K., Fairbank, J. *et al.* (2004). Fluoxetine, cognitive-behavioral therapy, and their combination for adolescents with depression: Treatment for adolescents with depression study (TADS) randomized controlled trial. *JAMA* 292(7), 807–820.

57 Wagner, K.D., Jonas, J., Findling, R.L., Ventura, D., and Saikali, K. (2006). A double-blind, randomized, placebo-controlled trial of escitalopram in the treatment of pediatric depression. *J. Am. Acad. Child Adolesc. Psychiatr.* 45(3), 280–288.

58 March, J.S., Silva, S., Petrycki, S., Curry, J., Wells, K., Fairbank, J. *et al.* (2007). The treatment for adolescents with depression study (TADS): Long-term effectiveness and safety outcomes. *Arch. Gen. Psychiatr.* 64(10), 1132–1143.

59 Goodyer, I., Dubicka, B., Wilkinson, P., Kelvin, R., Roberts, C., Byford, S. *et al.* (2007). Selective serotonin reuptake inhibitors (SSRIs) and routine specialist care with and without cognitive behaviour therapy in adolescents with major depression: Randomised controlled trial. *Br. Med. J.* 335(7611), 142.

60 Emslie, G., Kratochvil, C., Vitiello, B., Silva, S., Mayes, T., McNulty, S. *et al.* (2006). Treatment for adolescents with depression study (TADS): Safety results. *J. Am. Acad. Child Adolesc. Psychiat.* 45(12), 1440–1455.
61 March, J., Silva, S., and Vitiello, B. (2006). The treatment for adolescents with depression study (TADS): Methods and message at 12 weeks. *J. Am. Acad. Child Adolesc. Psychiatr.* 45(12), 1393–1403.
62 Hodes, M. and Garralda, E. (2007). NICE guidelines on depression in children and young people: Not always following the evidence. *Psychiatr. Bull.* 31, 361–362.
63 Watanabe, N., Hunot, V., Omori, I.M., Churchill, R., and Furukawa, T.A. (2007). Psychotherapy for depression among children and adolescents: A systematic review. *Acta. Psychiatr. Scand.* 116(2), 84–95.
64 McCarty, C.A. and Weisz, J.R. (2007). Effects of psychotherapy for depression in children and adolescents: What we can (and can't) learn from meta-analysis and component profiling. *J. Am. Acad. Child Adolesc. Psychiatr.* 46(7), 879–886.
65 Kaizar, E.E., Greenhouse, J.B., Seltman, H., and Kelleher, K. (2006). Do antidepressants cause suicidality in children? A Bayesian meta-analysis. *Clin. Trials* 3(2), 73–90. discussion 91–78.
66 Bridge, J.A., Salary, C.B., Birmaher, B., Asare, A.G., and Brent, D.A. (2005). The risks and benefits of antidepressant treatment for youth depression. *Ann. Med.* 37(6), 404–412.
67 Apter, A. and King, R.A. (2006). Management of the depressed, suicidal child or adolescent. *Child Adolesc. Psychiatr. Clin. N. Am.* 15(4), 999–1013.
68 Goldenberg, M.J. (2006). On evidence and evidence-based medicine: lessons from the philosophy of science. *Soc. Sci. Med.* 62(11), 2621–2632.
69 Cochrane, A. (1971). *Effectiveness and Efficiency: Random Reflections on Health Services.* Royal Society of Medicine Press, Cambridge.
70 Gibbons, R.D., Brown, C.H., Hur, K., Marcus, S.M., Bhaumik, D.K., Erkens, J.A. *et al.* (2007). Early evidence on the effects of regulators' suicidality warnings on SSRI prescriptions and suicide in children and adolescents. *Am. J. Psychiatr.* 164(9), 1356–1363.
71 Emslie, G.J., Wagner, K.D., Kutcher, S., Krulewicz, S., Fong, R., Carpenter, D.J. *et al.* (2006). Paroxetine treatment in children and adolescents with major depressive disorder: a randomized, multicenter, double-blind, placebo-controlled trial. *J. Am. Acad. Child Adolesc. Psychiatr.* 45(6), 709–719.

6 Media Coverage of Stem Cell Research

1 Nelkin, D. (1987). *Selling Science: How the Press Covers Science and Technology.* W.H. Freeman and Company, New York.
2 Nisbet, M.C., Brossard, D., and Kroepsch, A. (2003). Framing Science: The Stem Cell Controversy in an Age of Press/Politics. *The Harvard International Journal of Press/Politics* 8, 36–70.
3 Klimanskaya, I., Chung, Y., Becker, S.J. and Lanza, R.C. (2006). *Nature* 444, 481–485.
4 www.advancedcell.com, August 23, 2006
5 www.advancecell.com Press release for 6/21/2007

6 Alison Abbot. *'Ethical' stem cell paper under attack, Nature* 443, 12 (September 2006) doi: 10, 1038/443012a.
7 *Nature* doi: 10.1038/nature 05366 and doi: 10.1038/nature05608
8 *The LA Times* 8_24_06.pdf.
9 Nicholas Wade (www.nytimes.com/2006/08/23/science/23cnd) (archive).
10 Rick Weiss (www.washingtonpost.com/wp-dyn/content/article/2006/08/23/IAR2006082300936).
11 J.L. Simpson. *Blastomeres and stem cells Nature* 444, 432–435.
12 www.the-scientist.com/news/display/2463 and www.the-scientist.com/blog/display/24413
13 www.advancecell.com.
14 Kerr, D. *et al.* (2006). Recovery from paralysis in adult rats using embryonic stem cells. *Annals of Neurology* June, 32.
15 Lauran, N. (June 20, 2006) *Stem Cell Cocktail Helps Repair Paralysis in Rats* Associated Press, Washington.
16 Miranda, H. *Stem Cells May Help Reverse Paralysis* June 21, 2006 (www.webmd.com/braom/news/2000021/stem-cells-help-reverse.com).
17 Majorie, C. (June 26, 2006). *Embryonic Stem Cells Awaken Latent Nerve Repair.* Johns Hopkins University Gazette.
18 CBS News Transcripts. (June 21, 2006) *The Early Show.*
19 Majorie, C. (June 26, 2006). *Embryonic Stem Cells Awaken Latent Nerve Repair.* Johns Hopkins University Gazette (as picked up by HT Media).
20 Neurons Grown from Embryonic Stem Cells Restore Function in Paralyzed Rats in Mental Health Law Weekly July 15, 2006
21 Lauran, N. (June 20, 2006). *Stem Cell Cocktail Helps Repair Paralysis in Rats.* Associcated Press Worldstream.
22 Natalie, F. (June 20, 2006). *Neurons Grown from Embryonic Stem Cells Restore Function in Paralyzed Rats* NIH News.
23 Oscar, O. (June 22, 2006). *Stem Cell Therapy Restores Walking Function to Paralyzed Rats.* Medpage Today.
24 Schwartz, L., Woloshin, S., and Baczek, L. (2002). Media coverage of scientific meetings. Too much, too soon? *JAMA* 287, 2859–2863.
25 Nisbet, M.C., Brossard, D., and Kroepsch, A. (2003). Framing Science: The Stem Cell Controversy in an Age of Press/Politics. *The Harvard International Journal of Press/Politics* 8, 36–70.
26 Entwistle, V. (1995). Reporting research in medical journals and newspapers. *Br. Med. J.* 310, 920–923.
27 de Semir, V., Ribas, c., and Revuelta, G. (1998). Press releases of science journal articles and subsequent newspaper stories on the same topic. *JAMA* 280, 294–295.
28 Altman, L.A. (1995). The doctor's world; promises or miracles: News releases go where journals fear to tread. *The New York Times* 10 January, B6.
29 Woloshin, S. and Schwartz, L. (2002). Press releases. Translating research into news. *JAMA* 287, 2856–2858.
30 The Business Lawyer. (1996). 51, pp. 354–357.
31 Rieman, W., Rosen, R.A., Rosenfeld, S.B., and Smith, R.S. (1996). The Private Securities Litigation Reform Act of 1995; A User's Guide. *Securities Regulation Law Journal* 24, 118–122.

7 Medicine's Obsession with Disclosure of Financial Conflicts

1 Fontanarosa, P.B., Flanigan, A., and DeAngelis, C.D. (2005). Reporting conflicts of interest, financial aspects of research, and role of sponsors in funded studies. *JAMA* 294, 110–111.
2 Armstrong, D. (2006). Drug interactions: financial ties to industry cloud major depression study. *Wall Street Journal* July 11, A1.
3 Goldberg, C. (2006). Some seek to lift veil on research funding: full disclosure urged on money sources. *Boston Globe* August 8.
4 Armstrong, D. (2006). JAMA to toughen rules on author disclosure. *Wall Street Journal* July 12.
5 Peterson, M. (2003). Undisclosed financial ties prompt reproval of doctor. *New York Times* August 3.
6 Armstrong, D. (2006). Medical reviews face criticism over lapses. *Wall Street Journal* July 9.
7 Pincock, S. (2006). Journal editor quits in conflict scandal. *The Scientist* August 28.
8 Armstrong, D. (2005). Delicate operation: how a famed hospital invests in a device it uses and promotes. *Wall Street Journal* December 12.
9 Mauldin, C. (2006) (http://www.cspinet.org/new/pdf/integsci_3.31.pdf) March 31.
10 Willman, D. (2006). NIH audit criticizes scientist's dealings. *Los Angeles Times* September 10.
11 Saul, S. (2006). Ex-F.D.A. chief is charged with conflict. *New York Times, October* 17.
12 Dana, J. and Loewenstein, G. (2003). A social science perspective on gifts to physicians from industry. *JAMA* 290, 252–255.
13 Kassirer, J.P. (2004). *On the Take: How Medicine's Complicity with Big Business Can Endanger Your Health.* Oxford University Press, New York.
14 Weintraub, A. and Barrett, A. (2006). Medicine in conflict. *Business Week Online* 23. http://www.businessweek.com/magazine/content/06_43/b4006081.htm?chan = search, Oct.
15 Popp, R.L. and Smith, S.C. (2004). ACCF/AHA consensus conference report on professionalism and ethics. *J. Am. Coll. Card.* 44, 1722–1761.
16 Anonymous Program (2006). American Psychiatric Association. From science to public policy. Annual Meeting, Toronto, Canada, May 20–25.
17 Kassirer, J.P. (2004). Why should we swallow what these studies say? Washington Post. *Outlook* August 1.
18 Steinbrook, R. (2005). The controversy over Guidant's implantable defibrillators. *N Engl J Med* 353, 221–224.
19 M. Goozner (2007). Joint letter protesting conflicts of interest and imbalance at the Feb. 20 NIAD conference on neonatal herpes, Submitted to the NIH and Congress January 18, 2007.
20 Eichacker, P.Q., Natanson, C., and Danner, R.L. (2006). Surviving sepsis – practice guidelines, marketing campaigns, and Eli Lilly. *N. Engl. J. Med.* 355, 1640–1642.
21 Arshinoff, S. (2005). Excluding the experts? *CMAJ* , 173–849.
22 Bakalar, N. (2005). Potential conflicts cited in process for new drugs. *New York Times* October 25.

23 Jacobs, P. (2006). Federal rules often unenforced; schools are left to police themselves. *Mercury News* July 9.

24 Wade, N. (2006). *Before the Dawn: Recovering the Lost History of Our Ancestors.* Penguin Press, New York.

25 Cialdini, R.B. (1993). *Influence: Science and Practice.* Harper Collins College Publishers, New York.

26 Wang, S.S. (2006). Simply disclosing funds behind studies may not erase bias. *Wall St. Journal August* August 4, A11.

27 Wazana, A. (2000). Physicians and the pharmaceutical industry: Is a gift ever just a gift? *JAMA* 283, 373–380.

28 Katz, D., Caplan, A.L., and Merz, J.F. (2003). All gifts large and small: Toward an understanding of the ethics of pharmaceutical industry gift-giving. *Am. J. Bioeth.* 3, 39–46.

29 Stossel, T.P. (2007). Regulation of financial conflicts of interesting medical practice and medical research: A damaging solution in search of a problem. *Perspect Biol Med* 50, 54–71.

30 Stark, A. (2000). *Conflict of Interest in American Public Life.* Harvard University Press, Cambridge, MA.

31 Anonymous. (2006). Dealing with disclosure. *Nature Medicine* 12, 979.

32 Kassirer, J.P. (2006). A cure for public distrust. *Boston Globe* July 27, A11.

33 Laffin, M.B. (2005). Disclosing conflicts of interest can have damaging effects. *Carnegie Mellon press release* January 6.

34 Cain, D.M., Loewenstein, G., and Moore, D.A. (2005). The dirt on coming clean: Perverse effects of disclosing conflicts of interest. *J. Legal Studies* 34, 1–25.

35 Surowiecki, J. (2002). The talking cure. *The New Yorker. December* December 9.

36 Brennan, T.A., Rothman, D.J., Blank, L., Blumenthal, D., Chimonas, S., Cohen, J.J., Goldman, J., Kassirer, J.P., Kimball, H., Naughton, J., and Smelser, N. (2006). Health industry practices that create conflicts of interest: A policy proposal for academic medical centers. *JAMA* 295, 429–433.

37 Narins, R.G. and Bennett, W.M. (2007). Patient care guidelines: Problems and solutions. *Clin. J. Am. Soc. Nephrol.* 2, 1–2.

38 Steinbrook, R. (2007). Guidance for guidelines. *N. Engl. J. Med.* 356, 331–333.

39 Van Wyck, D., Eckardt, K.-U., Uhlig, K., Rocco, M., and Levin, A. (2007). Response to "Influence of industry on Renal Guideline Development". *Clin. J. Am. Soc. Nephrol.* 2, 13–14.

40 Anonymous. (2003). *Ethical journalism: Code of Conduct for the News and Editorial Departments.* New York Times, New York.

41 Thomas, L. Jr. (2007). Questions grow about a top CNBC anchor. *New York Times. February* February 12.

42 Rodwin, M.A. (1989). Physicians' conflicts of interest: the limitations of disclosure. *N. Engl. J. Med.* 321, 1405–1408.

43 Rodwin, M.A. (1993). *Medicine, Money, and Morals: Physicians' Conflicts of Interest* p. 219. Oxford University press, New York.

8 In Support of Industry-Sponsored Clinical Research

1 Thadeis, P. (2007). Pharma (http://www.phrma.org/innovation/).

2 Parexel Inc. (2005/2006). R&D Statistical Sourcebook.

3 CMR (2005/2006). International Factbook.
4 Tufts Impact Report (2006). Vol. 8, Issue 6, Nov/Dec.
5 US Government Accounting Office (November, 2006). www.gao.gov/newitems

9 Prevailing Truth: The Interface Between Religion and Science

1 For a useful introduction to the long-standing 'science and religion'. In *Fifty Years in Science and Religion: Ian G. Barbour and his Legacy*. (R.J. Russell, ed.). Ashgate, Aldershot.
2 So far as I am aware, 'media saturated' is my term to describe with as little judgement as possible the context in which we today live. See Carr, A.W. (1990). *Ministry and the Media*. SPCK, London. Also Tilby, A. (2004)' Spirituality and the Media' in Percy, M. and Lowe, S. R. (eds).' *The Character of Wisdom. Essays in Honour of Wesley Carr*. Ashgate, Aldershot pp. 125–140.
3 Attributed to Heracleitus, fl. 513 BC.
4 There is a tendency to treat all Greek philosophers as scientists – a word unknown until the 19th century. It can be misleading. For a useful table showing how intellectual developments occurred in contexts, see "The Big Picture" (http://www.swan.ac.uk/grst/big.htm).
5 Finochiarro, M.A. (2005). *Retrying Galileo 1633–1992*. University of California Press, Berkeley.
6 While on reflection the argument from design for the existence of God is not very strong, its emotional strength from time to time tides it over and ensures its continuing popularity.
7 Theology was once 'Queen of the Sciences', although more recently that position has been claimed by mathematics.
8 It was coined from the Latin scientia by an English philosopher William Whewell (1794–1866) on analogy with 'artist'.
9 Browning, R. (1802). Bishop Blougram's Apology. 'Apology' is used in a theological sense (i.e. defence). This robust poem, as with other work by Browning, both welcomes the present and future but regrets the loss of the past.
10 Lasch, C. (1987). *The Culture of Narcissism: American Life in an Age of Diminishing Expectations*. Abacus, London.
11 Darwin is buried in Westminster Abbey. This was a cause of much curiosity and, we might say, anxiety on the part of groups of Russian sailors who visited the Abbey when their ships docked at Tilbury. Having been told that the survival of the fittest, extracted from its context, was Darwin's belief and that he had forever destroyed Christianity with his understanding of evolution, they could not understand why he was there. Those who know better recall that it was with no sense of triumph or wish to undermine faith that Darwin published his work.
12 I have Parkinson's disease.
13 Midgley, M. (2002). *Evolution as a Religion*. Routledge, London; ibid. (2003). *The Myths We Live By*. Routledge, London. The word "myth" is used in a particular sense.
14 This is the stage at which the influence of the media begins to come to bear.
15 See Introductory chapter for this volume, authored by Professors Snyder and Mayes.

16 Myth" is used in a technical sense to refer to an underlying story through which individuals and groups acquire and preserve their history and so generate social cohesion. "(Myths are) symbolic narratives which explain the origins of the tribe and cosmos and which locate the values of a group or society in a large frame of reference – cosmic, historical or societal." "Myth" In Carr, A.W. (2002). *The New Dictionary of Pastoral Studies*. SPCK, London.
17 The running "battle" between Richard Dawkins and Alister McGrath seems to exemplify this closeness. See , Dawkins, R. (2006). *The God Delusion*. Bantam, London. McGrath, A. (2007). *The Dawkins' Delusion*. SPCK, London.
18 See below, p. 115, for discussion of the' Chamberlain case' which illuminates this point.
19 The word "irrational" is here used in a technical sense meaning "without conscious thought". At the time of the Princess of Wales's funeral I used it in this sense and was reported as "Dean calls mourners 'mad'".
20 Miller, E.J. (1993). *From Dependency to Autonomy: Studies in Organization and Change* p. 106. Free Association Press, London.
21 For a convincing demonstration that in many fields of study, notably mathematics and physical sciences, Arabs and others from the East were far ahead of West, see Sen, A. (2006). *Identity amd Violence. The Illusion of Destiny*. Harmondsworth, Penguin. It is the core of his book. But there still remains the question of how this advance on the whole failed to move into the West, so that European scientists had to 'rediscover' what had been known. That curiosity, whatever its historic origins, has firmly rooted 'scientific science' in the Jewish Christian tradition.
22 Arnold, M. (1867). *Dover Beach*.
23 The significance of imagination as defining human distinctiveness see Pannenberg, W. (1998). Kos, 152, pp. 42–45.
24 The implicit violence that runs both through Delgado and the design and execution of his demonstration has not been commented on.
25 Smith, R. (2006). Research misconduct: the poisoning of the wells. *J. Royal Soc Med*. 99, 237f.
26 Chadwick, O. (1966). *The Victorian Church, Part I* p. 514. A. and C. Black, London.
27 This was for many years formalised. The Novartis Foundation supported the Media Resource Service, the sole purpose of which was to maintain a directory of experts in the scientific fields who were willing and able to advise the press.
28 Goldsacre, Ben, (2005). Don't dumb me down, *The Guardian* September 8, 2005. The taxonomy of characteristics is also derived from this article.
29 My grandfather, William Cummins, owned "The Muswell Hill Record", a local newspaper and had no doubt that part of his responsibility was to inform and instruct. See Dick, A. (1943). *Inside Story*, Unwin, London.
30 Postman, N. (1986). *Amusing Ourselves to Death* p. 93. Methuen, London.

10 Science Meets Fundamentalist Religion

1 Kitzmiller *et al.* v. Dover Area School Board (2005). Plantiffs's exhibit P-124.
2 Kitzmiller *et al.* v. Dover Area School Board (2005). Baksa testimony 10/28/05: 53–56; Kitzmiller *et al.* v. Dover Area School Board (2005). Spahr testimony 10/12/05:72–73.

3 Kitzmiller*et al.* v. Dover School Board (2005). Buckingham testimony 10/27/05:36–37.
4 Kitzmiller *et al.* v. Dover School Board (2005). Buckingham testimony 10/27/05: 71–72.
5 Kitzmiller *et al.* v. Dover School Board (2005). Buckingham testimony 10/27/05: 50.
6 Kitzmiller *et al.* v. Dover School Board (2005). Buckingham testimony: 45.
7 United States Constitution, Amendment 1.
8 Everson v. Board of Education of Ewing Township (1947). 330 U.S.1, 15–16.
9 Kitzmiller *et al.* v. Dover School Board (2005). Plantiffs'exhibit P-124.
10 Kitzmiller *et al.* v. Dover School Board (2005). Alters testimony 10/12/05:112.
11 Kitzmiller *et al.* v. Dover School Board (2005). Alters testimony 10/12/05: 119.
12 Kitzmiller *et al.* v. Dover School Board (2005). Alters testimony 10/12/05: 110–111.
13 Kitzmiller *et al.* v. Dover School Board (2005). Forrest testimony 10/05/05: 103–105.
14 Kitzmiller *et al.* v. Dover School Board (2005). Forrest testimony 10/05/05: 91.
15 Edwards v Aguillard (1987). 482 U.S. 578: 598–599.
16 Kitzmiller *et al.* v. Dover School Board (2005). Forrest testimony 10/05/05: 123–124.
17 Kitzmiller *et al.* v. Dover School Board (2005). Buckingham testimony 10/27/05: 38–40.
18 Jones, J.E. III (2005). Kitzmiller *et al.* v. Dover School Board (2005), doc. 342, Memorandum of Opinion, 114.
19 Kitzmiller *et al.* v. Dover Area School Board (2005). Plantiffs' exhibit P-121.
20 Kitzmiller *et al.* v. Dover Area School Board (2005). Plantiffs' exhibit P-121.
21 Jones, J.E. III (2005). Kitzmiller *et al.* v. Dover School Board (2005), doc. 342, Memorandum of Opinion, 133–134.
22 McLean v. Arkansas (1982). 529 F Supp., 1267.
23 Jones, John III (2005). Kitzmiller *et al.* v. Dover School Board (2005), doc. 342, Memorandum of Opinion, 134.
24 Jones, John III (2005). Kitzmiller *et al.* v. Dover School Board (2005), doc. 342, Memorandum of Opinion, 131–132.
25 The text of the Lebec school district's Intersession course descriptions was transcribed by Nick Matzke of the National Center for Science Education from a faxed copy. It is available at: http://www.NCSEweb.org
26 As of 2000, the Center for Science and Culture (formerly the Center for the Renewal of Science and Culture) of the Discovery Institute posted a document known as the Wedge Strategy at: http://www.discovery.org/w3/discovery.org/crsc/aboutcrsc.html in these words. This document is no longer publicly accessible at that website but can be found at: http://www.antievolution.org/features/wedge.html. The Discovery Institute has acknowledged the authenticity of the document.
27 Crowther, R. (2006). Discovery Institute Statement to El Tejon School Board About Teaching Intelligent Design and Evolution (http://www.evolutionnews.org/2006/01/discovery_institute_statement.html).
28 Wanger, O.W. (2006). Hurst v. Newman *et al.* (2006), Stipulation for Dismissal with Prejudice, Order, 2.
29 Leshner, A. (2005). Debating How to Teach Science. Kansas City Star, May 8.
30 Information from http://www.ncseweb.org

31 Cicerone, R. (2005). Quoted in Kintisch, E. (2005). New National Academy Head Is No Stranger to Spotlight. *Science*, 309, 691; Fjerdingstad, E. (2005). A European Perspective on ID. *Science*, 309, 698.
32 Trends in International Mathematics and Science Study (2003). Report.
33 Information from http://www.ncseweb.org
34 Randerson, J. (2006). Revealed: rise of creationism in UK schools; PR packs spread controversial theory. The Guardian, November 27.
35 Enserink, M. (2005). Is Holland Becoming the Kansas of Europe? *Science* 308, 1394.
36 Graebsch, A. (2006). Polish scientists fight creationism. *Nature* 443, 890–891.
37 McKie, R. (2006). Kenya bishop leads anti-evolution fight; Evangelists want fossil exhibits kept out of sight. The Observer, August 10; Pflanz, M. (2006). Evangelicals urge museum to hide man's ancestors. *Telegraph*, August 12.
38 Sayin, U. and Kence, A. (1999). Islamic scientific creationism: A new challenge in Turkey. *Reports of the National Center for Science Education* 19(6), 18–20. 25–29.
39 Yahya, Harun (no date). Darwinism is not science. No publication given. Accessed 1/08/07 at: http://www.harunyahya.com/new_releases/news/shamanistic_religion.php
40 For a transcript of the debate "Evolution and Intelligent Design" between Nicholas Matzke and Mustafa Akyol on August 21, 2006 and September 3, 2006, see http://www.islamonline.net/livedialogue/english/Browse.asp?hGuestID = QYL0zb and http://www.islamonline.net/livedialogue/english/Browse.asp?hGuestID = a0iQ4n

11 Uneasy Alliance: The Intersection of Government Science, Politics and Policymaking

1 Gold, R.B. and Lehrman, L. (1989). Fetal research under fire: The influence of abortion politics. *Family Planning Perspectives* 21(1), 6–11. January-February (URL: http://www.jstor.org/view/00147354/di975908/97p0243m/0).
2 NIH Reauthorization, Hearings before the Subcommittee on Health and the Environment, U.S. House of Representatives. Ser. No. 102–24, April 15–16, 1991. p. 216. (Disclosure: Ruth J. Katz, author of this article, served as counsel to this Subcommittee, chaired by Congressman Henry Waxman, from 1982 to 1995.)
3 Pub. L. No. 110–5, The revised continuing appropriations resolution, 2007, February 15, 2007 (URL: http://www.govtrack.us/congress/bill.xpd?bill = hj110-20).
4 Gough, M. (2005). *Politicizing Science: The Alchemy of Policymaking* pp. 47–48. Hoover Institution Press, Stanford, CA.
5 Mooney, C. (2005). *Republican War on Science*p. 17. Basic Books, New York.
6 Union of Concerned Scientists. Surveys of scientists at federal agencies (URL: http://www.ucsusa.org/scientific_integrity/interference/survey-summaries.html).
7 Union of Concerned Scientists. Political interference in science (URL: http://www.ucsusa.org/scientific_integrity/interference).
8 Union of Concerned Scientists. Restoring scientific integrity in policymaking (URL: http://www.ucsusa.org/scientific_integrity/interference/scientists-signon-statement.html).

9 Integrity of Science Working Group. Science and integrity survey, September 21, 2004 (URL: http://www.ucsusa.org/assets/documents/scientific_integrity/Science_and_Integrity_Survey_1.pdf).

10 Allegations of Political Interference with the Work of Government Climate Change Scientists, Hearing before the Committee on Oversight and Government Reform, US House of Representatives, Statement of Roger A. Pielke Jr., Ser. No. 110–111, p. 96. Washington, DC: US Government Printing Office, 2007 (URL: http://oversight.house.gov/documents/20070130113315-90082.pdf).

11 Abramowitz, S. (1984). A Stalemate on Test-Tube Baby Research. *The Hastings Center Report* 14(1), 5–9. February (URL: http://www.jstor.org/view/00930334/ap060077/06a00090/0).

12 US Congress, Office of Technology Assessment. *Biomedical Ethics in US Public Policy-Background Paper,* OTA-BP-BBS-105. Washington, DC: US Government Printing Office, June 1993.

13 Department of Health, Education and Welfare, Ethics Advisory Board. (1979). *Report and Conclusions: HEW Support of Research Involving Human in vitro Fertilization and Embryo Transfer* (44 Fed Reg. 35,033, 35,056, June 18, 1979). US Government Printing Office, Washington, DC.

14 Wymelenberg, S. (1990). *Science and Babies: Private Decisions, Public Dilemmas*p. 12. National Academy Press, Washington, DC.

15 Centers for Disease Control and Prevention (1981), Pneumocystis Pneumonia – Los Angeles. *Morbidity and Mortality Weekly Report,* 30(21), 250–252.

16 In Their Own Words: NIH Researchers Recall the Early Years of AIDS (URL: http://history.nih.gov/NIHInOwnWords/docs/page_26.html).

17 Cimons, M. (1985). Health Experts Glad Reagan Cited AIDS; Pleased by Attention, Disappointed He Did Not Quell School Fear. *Los Angeles Times* September 19.

18 Smith, R.A. (1998). *The Encyclopedia of AIDS: A Social, Political, Cultural and Scientific Record of the HIV Epidemic* revised 2001. Penguin Putnam, New York.

19 Koop, C.E. (1991). *Koop: The Memoirs of America's Family Doctor.* Random House, New York.

20 Harris, G. (2007). Surgeon General Sees 4-Year Term as Compromised. *The New York Times* July 11.

21 Altman, D. (1987). *AIDS in the Mind of America*pp. 58–81. Anchor Books, New York.

22 Altman, D. AIDS in the Mind of America, p. 29, citing Talbot, D. and Bush. L., At Risk, Mother Jones, April 1985, p. 30.

23 Smith, P.W. and Swerdloff, J.T. (1988). Federal Funding for AIDS Research and Education. *Congressional Research Service* April 1.

24 Institute of Medicine. (1986). *Confronting AIDS: Directions for Public Health, Health Care, and Research* pp. 10–11. . National Academy Press, Washington, DC.

25 Department of Health and Human Services. Fetal Tissue Bank, Press release, May 19, 1992 (URL: http://www.hhs.gov/news/press/pre1995pres/920519.txt).

26 Hanna, K.E. (1991). *Biomedical Politics*p. 223. National Academy Press, Washington, DC.

27 PL 105–78, Department of Labor, Health and Human Services, and Education, and Related Agencies Appropriations Act, 1998. November 13, 1997

(URL: http://frwebgate.access.gpo.gov/cgi-bin/getdoc.cgi?dbname = 105_cong_public_ laws&docid = f:publ78.105.pdf).
28 HHS Press Office. Needle exchange programs: Part of a comprehensive HIV prevention strategy, April 20, 1998.
29 Interventions to Prevent HIV Risk Behaviors. National Institutes of Health's Consensus Development Conference Statement (URL: http://consensus.nih.gov/1997/1997PreventHIVRisk104html.htm).
30 Ban on Federal Funding of Needle Exchange Will Not Be Lifted, Says Clinton Administration. National drug strategy network, issue brief, March–April 1998 (URL: http://www.ndsn.org/marapr98/harm1.html).
31 Stolberg, S.G. (1998). Clinton Decides Not to Finance Needle Program. *The New York Times* April 21.
32 Shulman, S. (2006). *Undermining Science: Suppression and Distortion in the Bush Administration*. University of California Press, Berkeley, CA.
33 Mooney, C. (2005). *The Republican War on Science*. Basic Books, New York.
34 Politics and Science in the Bush Administration. Prepared for Rep. Henry, A. Waxman. Special Investigations Division, Committee on Oversight and Government Reform, US House of Representatives, November 13, 2003.
35 Specter, M. (2006). Political Science. *The New Yorker* March 13, 58–69.
36 Union of Concerned Scientists and US Government Accountability Project, Atmosphere of Pressure: Political Interference in Climate Change Science, February 2007 (URL: http://www.ucsusa.org/assets/documents/scientific_integrity/Atmosphere-of-Pressure.pdf).
37 Union of Concerned Scientists, Voices of Scientists at FDA: Protecting Public Health Depends on Independent Science, 2006 (URL: http://www.ucsusa.org/assets/documents/scientific_integrity/FDA-Survey-Brochure.pdf).
38 US Government Accountability Office, Federal Research: Policies Guiding the Dissemination of Scientific Research from Selected Agencies Should Be Clarified and Better Communicated, GAO-07-673, May 2007.
39 Broder, J.M. (2007). US Agency May Reverse 8 Decisions on Wildlife. *The New York Times* July 21.
40 Benson, E. (March 2003). Political Science. *Monitor Psychol.* 34(3). (URL: http://www.apa.org/monitor/mar03/political.html).
41 Union of Concerned Scientists, Scientific Integrity in Policymaking: An Investigation into the Bush Administration's Misuse of Science, March 2004 (URL: http://www.ucsusa.org/assets/documents/scientific_integrity/RSI_final_fullreport_1.pdf).
42 Committee on Science. (2005). *Engineering and Public Policy, Science and Technology in the National Interest: Ensuring the Best Presidential and Federal Advisory Committee Science and Technology Appointments* p. 10. National Academies Press, Washington, DC.
43 Kersten, D., HHS Monitors Scientists' Meetings with International Groups, Government Executive, August 2004 (URL: http://www.govexec.com/dailyfed/0804/080204dk1.htm).
44 Hadley. C. (2004). Science Policy in the USA. *European Molecular Biology Organization Report*, 5(10), 932–936.
45 Michaels, D. (June 2005). Doubt is their product. *Sci. Am.* 292(6).

46 Pear, R. (2007). Bush Directive Increases Sway on Regulation. *The New York Times* January 30.
47 Shulman, S. (2006). *Undermining Science: Suppression and Distortion in the Bush Administration*pp. 16–30. . University of California Press, Berkeley, CA.
48 Revkin, A.C. (2003). Report by the EPA Leaves Out Data on Climate Change. *The New York Times* June 19.
49 Hebert, H.J. (October 23, 2007). *White House Edits CDC Climate Testimony* URL: http://news.yahoo.com/s/ap/20071023/ap_on_sc/global_warming_health&printer=1;_ylt=Atx1kQWzXXZ6bTCYpfs0F9FxieAA]. Associated Press.
50 Melbye, M. *et al.* (January 9, 1997). Induced Abortion and the Risk of Breast Cancer. *New Eng. J. Med.* 336(2), 81–85. ((URL: https://content.nejm.org/cgi/content/abstract/336/2/81?ck=nck).
51 National Cancer Institute. *Summary Report: Early Reproductive Events and Breast Cancer*, March 25, 2003 (URL: http://www.cancer.gov/cancerinfo/ere-workshop-report).
52 Centers for Disease Control and Prevention. *Fact Sheet: Male Latex Condoms and Sexually Transmitted Disease* (URL: http://www.cdc.gov/condomeffectiveness/condoms.pdf).
53 Barr to Amend Age Restriction in Application for Nonprescription Sales of Plan B, Cannot Require Pharmacies to Abide by Law, CEO Says, *Daily Women's Health Policy Reports*, Kaisernetwork.org, August 10, 2006 (URL: http://www.kaisernetwork.org/daily_reports/rep_index.cfm?hint=2&DR_ID=39044).
54 Saul, S. (2006). FDA Shifts View on Next-Day Pill. *The New York Times* August 1.
55 Hall, S.S. (2003). Bush's political science. *The New York Times* June 12.
56 Rudoren, J. (2006). Stem cell work gets states' aid after Bush Veto. *The New York Times* July 25.
57 Arnold, W. (2006). Singapore acts as haven for stem cell research. *The New York Times* August 17.
58 Pielke, R.A. (2007). *The Honest Broker: Making Sense of Science in Policy and Politics*. Cambridge University Press, Cambridge, UK.
59 Pielke, R.A. (2007). *The Honest Broker: Making Sense of Science in Policy and Politics* p. 3. Cambridge University Press, Cambridge, UK.
60 Palca, J. (1994). A word to the wise. *The Hastings Center Report* 24(2). March.
61 Shilts, R. (1987). *And the Band Played On* p. 273. St Martin's Press, New York.
62 Shilts, R. (1987). *And the Band Played On* p. 295. St Martin's Press, New York.
63 Hebert, H.J. (2007). White House Chided for Editing Testimony, Associated Press, October 24, (URL: http://news.yahoo.com/s/ap/20071024/ap_on_go_ca_st_pe/global_warming_health).
64 Lee, C. (2007). Draft Reflects Tensions at HHS. *The Washington Post* July 31.
65 Allegations of Political Interference with the Work of Government Climate Change Scientists, Hearing before the Committee on Oversight and Government Reform, U.S. House of Representatives, Statement of Drew Shindell, Series. No. 110-1, p. 71. Washington, DC: U.S. Government Printing Office (2007). (URL: http://oversight.house.gov/documents/20070130113315-90082.pdf).
66 Piltz, R. (2005). On Issues of Concern about the Governance and Direction of the Climate Change Science Program. Memo. June 2 (URL: http://www.climatesciencewatch.org/index.php/csw/details/memo-to-ccsp-principals).

67 Rubin, R. (2007). FDA scientist says she was reprimanded for warning. *USA Today* June 10.
68 H.R. 985, 110th Congress (2007). Whisteblower Protection Enhancement Act of 2007.

12 On The Relations Between Scientists & Journalists: Reflections by an Ethicist

1 Snyder, P.J. and Mayes, L.C. Introduction to this book.
2 I was the NHRPAC member who responded to the bits of disinformation by supplying the relevant facts.
3 For further reading on this clinical trial, see: Fisher, D.G., Fenaughty, A.M., Cagle, H.H., and Wells, R.S. (2003). Needle exchange and injection drug use frequency: A randomized clinical trial. *J. Acquired Immune Deficiency Syndromes* 33(2), 199–205.
4 I served as chairperson of this committee.
5 I served as a consultant to the defense attorneys in each of these two cases.
6 For a detailed discussion of this case, see: Katz, J. (1993). Human Experimentation and Human Rights. *St Louis University Law J.* 38, 7–54.
7 For a discussion of this case see: Marshall, E. (2003). Fred Hutchinson Center under Fire. *Science* 292, 25. This case is also mentioned in the following review of litigation in the field of human subjects research: Mello, M.M., Studdert, D.M., and Brennan, T.A. (2003). The Rise of Litigation in Human Subjects Research. *Ann. Int. Med.* 139, 40–45.
8 The information in this paragraph is based on estimates made by persons in a position to know about such matters. As far as I know, the details have not been published. The exact amount of the settlement is protected by a confidentiality agreement. The main point I wish to make does not depend on the availability of exact numbers. The point is that cases of this type, which generate banner headlines when the plaintiffs' allegations are first made public, generate little or no journalistic coverage when the case is settled or otherwise resolved. The public, then, is left with the impression that something very bad went wrong at UCLA and this false impression is virtually never corrected in the media.
9 "Comparator" is the term used for the drug (or placebo) administered to members of the control group in a clinical trial. In this case, one of the compounds used as a comparator was naproxen, a drug marketed under several brand names including Aleve® and Naprosyn®. In this case I do not mean to support a position on the ethical propriety of withholding the information about the adverse effects of Vioxx®; rather, I just mean to illustrate how a single set of 'facts' can be interpreted very differently depending on one's experiences.
10 For a portal of entry into the literature on the value of making consent forms more readable by reducing the "grade level" of their content, see Levine, R.J. (1988). *Ethics and Regulation of Clinical Research.* Yale University Press, 2nd Edition, New Haven, pp.138–139. In this passage I explain why I don't consider this a reliable solution to solving the problem of increasing comprehension during the process of informed consent.

11 MacIntyre, A. (1984). *After Virtue* pp. 187ff. 2nd edition. University of Notre Dame Press, Notre Dame, Indiana.

12 There are some people who are not members of a practice who have acquired sufficient knowledge and understanding of the internal goods of the practice to comment fully on the activities of practitioners. Some have acquired this understanding by "submitting to the tutelage" of experienced practitioners for a time before they became journalists or ethicists by, for example, enrolling in medical schools or in graduate programs in relevant sciences. Others appear to me to have accomplished such understandings by joining teams of health professionals as they engage in their "work rounds" or by service on IRBs or hospital ethics committees. A full discussion of alternate routes to 'sufficient' understanding is beyond the scope of this chapter.

13 MacIntyre, p. 191.

14 Committee on the Conduct of Science. (1989). National Academy of Science: On Being a Scientist. *Proceedings of the National Academy of Sciences* 86, 9053–9074.

13 Don't Shoot the Messenger

1 Seventh report: Scientific Evidence, Risk and Evidence Based Policy Making. House of Commons Select Committee on Science and Technology (2006). London (http://www.publications.parliament.uk/pa/cm200506/cmselect/cmsctech/900/90002.htm).

2 Guidelines on science and health communication (2001). Royal Society. London (www.royalsoc.ac.uk/document.asp?tip = 0&id = 1412).

3 Towards a better map: Science, the public and the media. Economic and Social Research Council (2003). Swindon (http://www.esrc.ac.uk/ESRCInfoCentre/Images/Mapdocfinal_tcm6-5505.pdf).

4 Third report. House of Lords Select Committee on Science and Technology (2001). London (www.publications.parliament.uk/pa/ld199900/ldselect/ldsctech/38/3810.htm).

14 Future Trends in Medical Research Publishing

1 Dinsdale, J.E. (2001). Medical publishing: remembrance of things past and intimations of the future. *Psychosomat. Med.* 63, 1–6.

2 Solomon, D.J. (2002). Talking past each other: making sense of the debate over electronic publication. *First Monday* 7(8), 1–14.

3 Smith, R. (2006). *The trouble with medical journals*. Roy. Soc. Med., London.

4 Peer Review and BioMedical Publication. Peer Review Congress. June 2007. 12 September 2007 (http://www.ama-assn.org/public/peer/peerhome.htm).

5 Jefferson, T., Alderson, P., Wager, E., and Davidoff, F. (2002). Effects of editorial peer review: A systematic review. *JAMA* 287, 2784–2786.

6 Falagas, M. (2007). Peer Review in Open Access Journals. Open Medicine. 12 September (http://www.openmedicine.ca/article/viewArticle/35/22).

7 Peters, D. and Ceci, S. (1982). Peer-review practices of psychological journals: The fate of submitted articles, submitted again. *Behav. Brain Sci.* 5, 187–255. (cited by Smith 2006).

8 The Budapest Open Access Initiative (2007). Soros. Open Society Institute. 23 October 2007 (http://www.soros.org/openaccess/commitment.shtml).

9 General Public Liscense. GNU. June 29 2007. 23 October 2007 (http://www.gnu.org/copyleft/gpl.html).

10 Public Knowledge Project. PKP. July 11 2007. University of British Colombia. 23 October 2007 (http://pkp.sfu.ca/).

11 Digital Object Identifier (2007). Wikipedia. 23 October (http://en.wikipedia.org/wiki/Digital_object_identifier).

12 Wiki. What is Wiki. June 27 2007. 23 October 2007 (http://wiki.org/wiki.cgi?WhatIsWiki).

13 Real Simple Syndication (2007). Wikipedia. 23 October 2007 (http://en.wikipedia.org/wiki/RSS_(file_format)).

14 What is Web 2.0. O'Reilly. (2005). School of Technology. 23 October (http://www.oreilly.com/pub/a/oreilly/tim/news/2005/09/30/what-is-web-20.html).

15 Copyleft. Wikipedia. (2007). 23 October (http://en.wikipedia.org/wiki/Copyleft).

16 Denoyer, G. Wikipedia. The Wikipedia XML Corpus (2007). 12 September (http://citeseer.ist.psu.edu/krotzsch05wikipedia.html).

17 Suber, P. (2007). Faculty of 1000. *J. Biol.* 1.318 June 2002 1–3. 10 September 2007 (http://jbiol.com/content/pdf/1475-4924-1-1.pdf).

18 Walker; Holt. (2007). An exclusive interview with the creators of Philica. *Realy Magazine* 12 September 2007 (http://www.ohpurleese.com/interview.htm).

19 BioMed Central. (2007). *What is BioMed Central?* 12 September 2007 (http://www.biomedcentral.com/info/about/whatis).

20 PMC Overview. (2007). *PubMed Central.* PubMed Central. 12 September 2007 (http://www.pubmedcentral.nih.gov/about/intro.html).

21 Frequently asked questions about PLoS and PLoS Medicine. (2007). *PLoS Medicine* Public Library of Science. 12 September 2007 (http://journals.plos.org/plosmedicine/faq.php).

22 ArXive. (2007) *The Wikipedia XML Corpus.* 12 September 2007 (http://wikipedia.org/wiki/ArXiv).

23 Vine, R. (2006). Google Scholar. *PubMed Central,* 97–99. 10 September 2007 (http://www.pubmedcentral.nih.gov/articlerender.fcgi?artid = 1324783).

24 HINARI. (2007). *Health InterNewtwork Access to Research Initiative.* World Health Organization. 12 September 2007 (http://www.who.int/hinari/en/).

25 Squires, F. and Bruce, S. (2005). The World Association of Medical Editors (WAME): Thriving in its first decade. *Science Editor* 28, 13–16. http://www.councilscienceeditors.org/members/securedDocuments/v28n1p013-016.pdf.

26 Till, J. (2007). *Peer Review in a Post e-Prints World: A Proposal,*12 September (http://www.jmir.org/2000/3/e14).. . University Health Network.

27 Lorimer, R., Smith, R., and Wolstenholme, P. (2000). Fogo Island goes digital: taking a scholarly journal on line, the case of CJC-online.ca. *Can. J.Comm.* 25(3).

28 Till, J. and Leishman, J. (2006). Be openly accessible or be obscure? *University of Toronto Bulletin* 11 October.

29 Richardson, A. (2004). Future Trends in Medical Publishing. *Medizin-bibliothek-information* 4(20).

15 Conclusions Section

1 Coming Down Hard, First aired 10/6/2004 Session E5308 in Season 15; *Law and Order*.
2 Krimsky, S. (2003). *Science in the Private Interest: Has the Lure of Profits Corrupted Biomedical Research?*. Rowan and Littlefield Publishers, Inc., Lanham, MD.
3 Krimsky, S. (2003). A conflict of interest. *New Scientist* 179(2410), 21. 30 August.
4 Als-Nielsen, B., Chen, W., Gluud, C., and Kiaergard, L.L. (2003). Association of funding and conclusions in randomized drug trials: A reflection of treatment effect or adverse events? *J. Am.. Medical Assoc.* 290(7), 921–928.
5 Chalmers, I. (2004). In the dark. *New Scientist* 06 March, 19.
6 [anonymous editorial] When drug companies hide data (2004). *The New York Times*, 06 June (URL: http://query.nytimes.com/gst/fullpage.html?sec=health&res=9F01E4DC1031F935A35755C0A9629C8B63).
7 Kramer, T.A.M. (2004). Psychopharmacology in the New York Times. *Medscape Gen. Med.* 6(2). posted 23 June..
8 Hemmini, E. (1980). Study of information submitted by drug companies to licensing authorities. *BMJ* 280(6217), 833–836.
9 Antonuccio, D.O., Danton, W.G., and McClanahan, T.W. (2003). Psychology in the prescription era. Building a firewall between marketing and science. *Am. Psychol.* 58(12), 1028–1043.
10 Yank, V. and Rennie, D. (1999). Disclosure of researcher contributions: A study of original research article in The Lancet. *Ann. Int. Med.* 130(8), 661–670.
11 Bates, T., Anic, A., Marusic, M., and Marusic, A. (2004). Authorship criteria and disclosure of contributions: Comparison of 3 medical journals with different author contribution forms. *J. Am. Med. Assoc.* 292(1), 86–88.
12 Tufte, E.R. (1997). Visual and statistical thinking: Displays of evidence for making decisions (Chapter 2). *Visual Explanations*. Graphics Press, LLC., Cheshire, CT, pp. 39–53.
13 Biever, C. (2006). The God lab. *New Scientist* December 16, 8–11.
14 Nissen, S.E. and Wolski, K. (2007). Effect of rosiglitazone on the risk of myocardial infarction and death from cardiovascular causes. *N. Eng. J. Med.* 356(24), 2457–2471.
15 Gottlieb, S. (2007). Journalistic malpractice. *Editorial in the Wall Street Journal* May 29, A15.
16 Jiang, Y., Jahagirdar, B.N., Reinhardt, R.L., Schwartz, R.E., Keene, C.D., Ortiz-Gonzalez, X.R., Reyes, M., Lenvik, T., Lund, T., Blackstad, M., Du, J., Aldrich, S., Lisberg, A., Low, W.C., Largaespada, D.A., and Verfaillie, C.M. (2002). Pluripotency of mesenchymal stem cells derived from adult marrow. *Nature* 418, 41–49.
17 Aldhous, P. and Reich, E.S. (2007). Fresh questions on stem cell findings. *New Scientist* March 24–30, 12–13.
18 Reyes, M., Lund, T., Lenvik, T., Aguiar, D., Koodie, L., and Verfaillie, C.M. (2001). Purification and ex vivo expansion of postnatal human marrow mesodermal progenitor cells. *Blood* 98, 2615–2625.

19 Coghlan, A. (2004). Medical editors to take tougher line. *New Scientist* March 6, 11.
20 Craig, I.D., Plume, A.M., McVeigh, M.E., Pringle, J., and Armin, M. (2007). Do open access articles have greater criterion impact? A critical review of the literature. *J. Informatics* 1, 239–248.
21 Swan, A. (2007). Open access and the progress of science. *American Scientist* 95, 197–198.
22 Woodward, J. and Goodstein, D. (1996). Conduct, misconduct and the structure of science. *American Scientist* 84(5), 479–490.
23 Kitcher, P. (1990). The division of cognitive labor. *J. Philos.* 87, 5–22.

About the Senior Authors

Peter J. Snyder, Ph.D.

Vice President for Clinical Research, Rhode Island Hospital, The Miriam Hospital and Bradley
Hospital, Providence, RI, USA
Professor, Department of Clinical Neurosciences (Neurology), The Warren Alpert Medical School of Brown University,
Providence, RI, USA
Adjunct Professor, Child Study Center, Yale University School of Medicine, New Haven, CT, USA

Professor Snyder received his doctorate in clinical psychology and behavioral neuroscience from Michigan State University in 1992, following an internship in clinical neuropsychology at the Long Island Jewish Medical Center (New York, USA). Dr. Snyder was awarded the 1992 Wilder Penfield Fellowship by the American Epilepsy Society and the Epilepsy Foundation of America, and he served as a Clinical Neurosciences Fellow in the NIMH Clinical Research Center for the Study of Schizophrenia at Hillside Hospital (Albert Einstein College of Medicine) in 1992 and 1993.

Dr. Snyder publishes regularly and he maintains numerous scientific collaborations. He serves as an Associate Editor of *Brain and Cognition*, and he has delivered over 120 presentations at international scientific conferences. His academic interests span across a range of topics in neuropharmacology, neurophysiology, history of neuroscience and research ethics, and his clinical interests bridge a wide variety of neurological and neuropsychiatric conditions. From 1998 through 2005, Dr. Snyder was employed as a scientist, clinician and manager at Pfizer Global Research & Development – Groton Laboratories (Connecticut, USA) – the largest research laboratory of Pfizer Inc. Dr. Snyder was responsible for the identification and development of novel clinical technologies and biomarkers for the CNS therapeutic area at Pfizer. Then, as a Director and Early Clinical Leader at Pfizer, Dr. Snyder led the development of novel compounds for the treatment of schizophrenia and Alzheimer's disease.

Dr. Snyder left Pfizer in 2006, to return to academia as a Professor (with tenure) in the Departments of Psychology and Neurology at the University of Connecticut (Storrs and Farmington, Connecticut). Dr. Snyder directs a Master's program in Clinical and Translational Research, he teaches graduate courses in cognitive neuroscience and clinical neuropsychology, he maintains an active laboratory and he serves as the major professor for several graduate and post-doctoral students. Dr. Snyder consults to several pharmaceutical corporations, and he is an active adjunct faculty member in the Yale Child Study Center, Yale University School of Medicine.

Dr. Snyder is a Fellow of the American Psychological Association (Div. 40, Clinical Neuropsychology; Div. 6, Behavioural Neuroscience), and he is the recipient of the 2001 Distinguished Early Career Contributions Award from the National Academy of Neuropsychology. Dr. Snyder is the Senior Editor of a best-selling handbook on clinical neuropsychological practice that is published by the American Psychological Association (APA Books, Inc.).

Linda C. Mayes, M.D.
Arnold Gesell Professor of Child Psychiatry, Pediatrics, Psychology, and Epidemiology and Public Health,
Child Study Center, Yale University School of Medicine, New Haven, CT

Dr. Linda Mayes is the Arnold Gesell Professor of Child Psychiatry, Pediatrics, Psychology, and Epidemiology and Public Health in the Yale Child Study Center. She is also Special Advisor to the Dean in the Yale School of Medicine. After graduating from the University of the South in 1973, she received her medical degree at Vanderbilt University in 1977. Following an internship and residency in pediatrics, she spent 2 years as a fellow with in the division of neonatology and worked in the area of developmental outcome of high-risk preterm infants. Dr. Mayes's interest in the long-term impact of perinatal biological and psychosocial stressors developed during that fellowship and she came to Yale in 1982 to do a Robert Wood Johnson General Academic Pediatrics Fellowship. During this Fellowship she closely collaborated with faculty in the Department of Psychology and in the Child Study Center, and she joined the Center's faculty in 1985 to establish a laboratory for studying infant learning and attention. Subsequently, she also developed a neurophysiology laboratory for studies of the startle response and related indices of emotional regulation in children and adolescents and currently oversees the Developmental Electrophysiology Laboratory.

Dr. Mayes's research integrates perspectives from child development, behavioral neuroscience, psychophysiology and neurobiology, developmental psychopathology and neurobehavioral teratology. She has published widely in the developmental psychology, pediatrics and child psychiatry literature. Her work focuses on stress-response and regulatory mechanisms in young children at both biological and psychosocial risk. She has made contributions to understanding the mechanisms of effect of prenatal stimulant exposure on the ontogeny of arousal regulatory systems and the relation between dysfunctional emotional regulation and impaired prefrontal cortical function in young children.

Dr. Mayes is also trained as an adult and child psychoanalyst and is the Chairman of the Directorial Team of the Anna Freud Centre in London as well as the coordinator of the Anna Freud Centre program at the Yale Child Study Center. In this capacity, she focuses on developing research relevant to basic psychoanalytic theories of mental development as well as mentoring young scholars interested in the interface between psychoanalytic theory and developmental science. With her colleagues Peter Fonagy, Ph.D. and Mary Target, Ph.D. in London, she oversees a new masters program in psychodynamic developmental neuroscience offered collaboratively between University College London and the Yale School of Medicine. Dr. Mayes is a visiting professor at University College London where she participates regularly as a member of a research faculty training program and is also on the adjunct faculty of the Department of Psychology, at the University of Connecticut.

Dennis D. Spencer, M.D.
Harvey and Kate Cushing Professor of Neurosurgery
Chair, Department of Neurosurgery
Yale University School of Medicine, New Haven, CT, USA

Dr. Spencer is the Harvey and Kate Cushing Professor and Chair of the Department of Neurosurgery at Yale University School of Medicine. He is a graduate of Washington University School of Medicine and completed his neurosurgical residency at Yale in 1977. He joined the Yale neurosurgery faculty following his residency, and became Chief of Neurosurgery in 1987. He has an international reputation in the surgical treatment of neurological diseases causing epilepsy and developed a widely used surgical approach designed to spare the excision of as much neocortex as possible, for patients with temporal lobe epilepsy.

His research has brought together basic scientists and clinicians around a program concerning energetics, glutamate metabolism and the neurobiological study of human epileptogenic tissue. Study techniques include 4T MRS, C13 intraoperative glucose turnover studies, and *in vivo* and *in vitro* electrophysiology and microdialysis, immunohistochemistry, confocal and EM microscopy, and molecular biology. In particular, laboratory discoveries are correlated with the epileptogenic substrate in order to help define human epilepsy pathogenesis and potential therapies.

Dr. Spencer was the 1999 recipient of the American Epilepsy Society's Research Award in Clinical Investigation, and the 2006 Society of Neurological Surgeons' Grass Award for Excellence in Research. He is past Chairman of the American Board of Neurological Surgery, Vice Chairman of the Neurosurgery Residency Review Committee, incumbent President of both the American Epilepsy Society (2008) and the Society of Neurological Surgeons, and in 2003-2004 he served as interim Dean of the Yale University School of Medicine.

Index

J

K

L

M